ASQ's Pocket Guide Series

REMOTE AUDITING FUNDAMENTALS

Second Edition

Shauna Wilson
Paul Russell

QUALITY PRESS

Milwaukee, Wisconsin

American Society for Quality, Quality Press, Milwaukee 53203
© 2024 by Quality Press
All rights reserved. Published 2024
Printed in the United States of America

28 27 26 25 24 {LS} 5 4 3 2 1

Publisher's Cataloging-in-Publication data

Names: WIlson, Shauna M, 1959–, author. | Russell, Paul B, 1974–, author.
Title: Remote auditing fundamentals , second edition / by Shauna Wilson
    and Paul Russell.
Description: Milwaukee, WI: Quality Press, 2024.
Identifiers: LCCN: 2024946766 | ISBN: 9781636941943 (paperback) |
    9781636941950 (pdf) | 9781636941967 (epub)
Subjects: LCSH Auditing--Data processing. | Auditing. Internal. |
    Auditing. | BISAC BUSINESS & ECONOMICS / Auditing
Classification: LCC HF5667.12 .R87 2024 | DDC 657/.458—dc23

ASQ advances individual, organizational, and community excellence
worldwide through learning, quality improvement, and knowledge exchange.

Bookstores, wholesalers, schools, libraries, businesses, and organizations:
Quality Press books are available at quantity discounts for bulk purchases
for business, trade, or educational uses. For more information, please contact
Quality Press at 800-248-1946 or ask@asq.org.

To place orders or browse the selection of all Quality Press titles,
visit our website at: http://www.asq.org/quality-press.

QUALITY
PRESS
Quality Press
600 N. Plankinton Ave.
Milwaukee, WI 53203-2914
Email: books@asq.org
Excellence Through Quality™

Dedication

The late J.P. Russell co-authored the original edition with Shauna Wilson. He was president of J.P. Russell & Associates in Gulf Breeze, Florida. He also was managing director of QualityWBT Center for Education, which provides online training (www.QualityWBT.org). Russell was an ASQ fellow and an ASQ-certified quality auditor (CQA), and former RAB Lead Auditor. He was the second recipient of the Paul Gauthier Award, the ASQ Audit Division's highest recognition award (Paul Gauthier was the first, and Shauna was the fourth recipient. There have only been four recipients awarded this honor since its creation in 2007). Russell was a voting member of the American National Standards Institute/ASQ Z1 committee, and member of the U.S. technical advisory group for the International Organization for Standardization (ISO) technical committee 176 (quality) and PC 302 (auditing). He was the editor of *The ASQ Auditing Handbook*, First through Third Editions, columnist for *Quality Progress* magazine, and author of several ASQ Quality Press books, including *The Process Auditing and Techniques Guide,* Second Edition. He was known by many of his ASQ colleagues as a professional who had a passion for quality auditing.

Contents

Contents

List of Figures and Tables

Preface

The purpose of this book is to provide hands-on guidelines for using electronic communication tools as part of the auditing process.

The pros and cons of conducting remote audits and their consequences will be reviewed. There are situations when remote auditing techniques are more efficient, and other times they may be less efficient and even lead to questionable audit report conclusions. In this book, we provide proven techniques for conducting remote audits and explore remote auditing practices to help organizations make informed decisions regarding their use.

Preface

Chapter 1

Introduction to the
Remote Auditing Process

I dentify environmental driving forces, issues, and important terms related to our virtual world:

- Characterize the virtual environment, telecommuter workforce, and remote auditing
- Review environmental driving forces
- Identify current challenges while working remotely
- Identify fundamental components to be successful

VIRTUAL ORGANIZATION AND THE TELECOMMUTER

A virtual organization employs a workforce that conducts business across time zones, geographic borders, and cultures. Another dimension of a virtual organization is that it can include partnerships with other companies, even competitors, for purchasing OEM parts or for contracting business functions like human resources or internal auditing from subject matter experts. Virtual organizations can be ongoing or only temporary. Temporary virtual organizations may disband when the project is complete, like a rock and roll concert tour or an election staff. Companies, however, now depend on the flexibility, increased productivity, and agility virtual organizations offer, along with a reduction in absenteeism, overhead and travel expenses, and improved morale. It looks like virtual organizations are here to stay.

According to a Pew Research Center survey, about 35% of workers with jobs that can be done remotely are working from home all the time and 41% of those with jobs that can be done remotely are working a hybrid schedule.[1] These statistics reflect the continued state of the virtual working environment. A telecommuter conducts work from a home, a telework center, or another location other than the central office. When auditing different companies, you will find that each has its telecommuter description; Table 1.1 shows a few commonly known telecommuter types.

Table 1.1   Virtual workforce.

Type of virtual worker	Description
Home telecommuter	Part-time one or two days per week or full-time
Telecenter/satellite office telecommuter/hoteling	Part-time one or two days per week in a remote facility or full-time
Virtual office worker	No office assigned
Long-distance telecommuter	Home worker in a distant city or state
Mobile professionals	May have office, but often on the road
Independent home worker	Self-employed, contract workers, lone eagles
Remote field worker	Full-time in the field, usually in a defined region
Decentralized work groups	People in the same group but in many locations
Remote branch office worker	Entire work group in a remote location

AUDITING REMOTELY

When you start talking about remote auditing with a group of auditors, you will hear plenty of strong opinions. Some auditors are vehemently opposed to the practice, while others are open to the idea (J.P. Russell,

---

[1] Kim Parker, Pew Research Center, "About a third of U.S. workers who can work from home now do so all the time," www.pewresearch.org, https://www.pewresearch.org/short-reads/2023/03/30/about-a-third-of-us-workers-who-can-work-from-home-do-so-all-the-time/ (accessed July 30, 2024).

"Auditing in Virtual Environments," *Quality Progress*, January 2011). The obvious benefit of remote auditing is a more efficient use of resources. Remote auditing techniques can save auditor travel time and expenses while improving efficiency (J.P. Russell, "Auditing in Virtual Environments"). When doing remote audits, practitioners have said they understand the business systems better than when conducting face-to-face audits, especially when the on-site audit involves being taken to a conference room to review binder after binder of procedures and records.

Generally, remote auditing has the potential to assist in many ways, such as helping companies solve their online communication issues. If communication issues stem from limited training for remote workers, a remote audit would have exposed this gap. Another benefit of remote audits could result from replicating the new emerging online workforce's environment. Auditors would become more familiar with how to communicate virtually to better understand the communication issues affecting these online management systems. Remote auditing could help to streamline and maintain websites, documentation, and record control in the online environment through scheduled reviews. Finally, remote auditing is a socially responsible alternative that reduces the consumption of energy and other resources.

IS/IS NOT

Understanding what remote auditing is and is not (see Table 1.2) will explain more about this emerging auditing technique. Most people don't know what to expect.

DRIVING FORCES

Within a short time, we've witnessed major events that permanently impacted the way business is conducted. Post-9/11 affected not only the way we travel but the economics of travel. COVID-19 changed our business relations in the international marketplace and the workplace environment to a virtual from a face-to-face (FtF) model. Technology continues to change the way we communicate.

Table 1.2   What remote auditing is and is not.

Is	Is not

• Uses the same audit process steps:
  – Prepare
  – Perform
  – Report
  – Follow up
	• A desk audit in which the auditee sends all records and procedures to an auditor to review in isolation.

• Simultaneous:
  – An auditor interviews the auditee in real time.
  – An auditor uses a computer, the internet, and cameras to view records and processes at the location.
	• An unattended auditor reviewing a company's computer files.
	
	• An electronic survey or checklist of questions to answer.
	
	• Records emailed to an auditor to review at a later date.

• Includes system assessments such as record control and procedure and process review.	
• The auditee has control of their online record system throughout the audit.	

Economic Impact

Travel is extremely expensive both in lost billable time and unavoidable airline delays and travel expenses. For some, it takes a minimum of six hours to travel, not including the flying time, because they live two hours from a major airport. Labor and lost income are hidden productivity costs, and in many cases are the larger costs.

Figure 1.1 shows a list of potential expenses and lost income. Can you think of other losses for your situation?

After allowing for travel to and from a location, you could calculate 12 lost billable hours at an inclusive rate (hourly plus overhead) of $150 an hour, which totals $1800, or one audit day of possible income. After calculating the added expenses of at least $1200 per day, it quickly totals to $3000 for a day of travel. If an auditor spends half or more of their time traveling by plane or on the road, remote auditing offers savings of half a year in labor plus associated travel expenses.

Lost income	Expenses
■ Movement to airport—1 hour	■ $600 airfare (two-hour flight)
■ Wait at airport—2 hours	■ $300 hotel
■ Flying time—1 hour	■ $200 rental car
■ Movement to rental car—1 hour	■ $100 food
■ Movement to client site—1 hour	

Figure 1.1 Potential travel expenses and lost income.

Other economic impacts include the immediate access to distal locations remote auditing provides. On an as-needed basis, remote auditing can be used in supplier audits. Auditors are also able to audit longer because the report writing is greatly reduced when an auditor can keyboard and record audit trails simultaneously.

International Marketplace

In the last 85 years, the television set has been one of the key drivers of globalization. Rather than having TV manufacturers, TV parts are developed and standardized by original equipment manufacturers (OEMs) and then shipped to a contract manufacturer, where the TV is assembled and shipped to retailers. As companies move to lower-cost countries, the price of TV sets falls. The trick is no longer who produces the best TV, but which company can create an effective and efficient global supply chain. The global marketplace has made virtual communication a way of life for many companies, which is changing the way auditors need to do their jobs.

> *Auditors may have more time to audit during remote audits because report-writing time may be reduced.*

Global trade is not a new phenomenon. In fact, it dates to 1450, providing countries access to raw materials and manufactured products (Vale Center for the Study of Globalization). Figure 1.2 identifies key events leading us to global trade.

The International Organization for Standardization (ISO) is the world's largest standards-developing organization. Since 1947, ISO has generated more than 24,600 standards. ISO has published four standards that are useful to reference when conducting remote audits: ISO 9001 for quality systems, ISO 27001 for information security management, ISO 31001 for risk management, and ISO 26000 for social responsibility.

These standards provide good advice for auditing global supply chains. Governments such as the United States are deferring to the use of international standards when they are available instead of maintaining a national standard that addresses the same subject.

Technology

Beginning in the early 1970s, technology advanced much more quickly than most of the working population could keep pace with. In 2001, collaborative programs were very expensive and exclusive compared to the solutions we have today. Affordable collaborative solutions now work on most mobile devices, like iPads or smartphones, and are available for small to large businesses (see Figure 1.3). Mobile device cameras make it very easy to see processes at remote locations.

In the past, machines served us, but the new technology of today is more like a partner. Interfacing with innovative technology takes some adjustment on our part, and not everyone wants to do that.

Many auditors, especially certification/registration auditors who have missed out on the technical advancements in working virtually, will be hesitant about the notion of remote auditing. Many tried but failed to conduct an online audit satisfactorily. Their stories not only tell of technology issues, but also of their frustration about how to engage with unseen people at the other end of the phone line. Plus, our new technology partner does not always work when we want it to and has its own lingo that we must learn.

Trade/conquest	New ideas create new markets	A small world emerges when technology advances	63,000 corporations— globalize
Trade, conquest, religion, and adventure lead to the development of new markets.	Civil war prompts creation of canned goods. Ford needs rubber for tires for its Model T car. Malaysia supplies rubber.	Collapsed distances and lowered transport costs Steam engine/ship Airplane Telegraph/telephone Radio Television Satellite Internet	Trade agreements (GATT, WTO, NAFTA) Offshore locations reduce manufacturing costs, shifting jobs across borders Billions of consumers and tourists

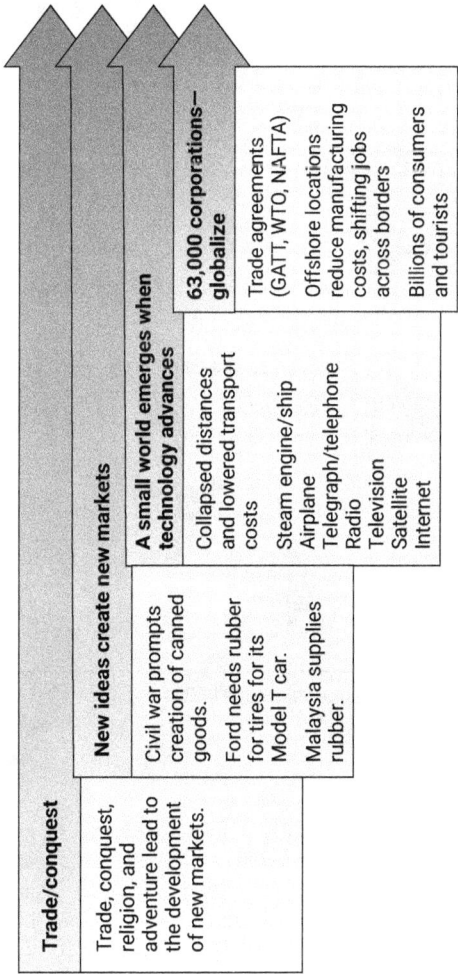

Figure 1.2 Key events leading to global trade.

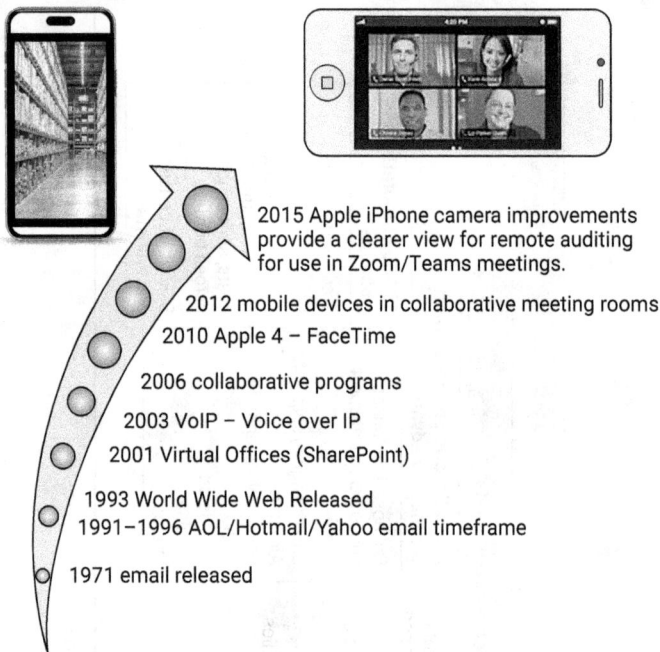

2015 Apple iPhone camera improvements provide a clearer view for remote auditing for use in Zoom/Teams meetings.

2012 mobile devices in collaborative meeting rooms

2010 Apple 4 – FaceTime

2006 collaborative programs

2003 VoIP – Voice over IP

2001 Virtual Offices (SharePoint)

1993 World Wide Web Released

1991–1996 AOL/Hotmail/Yahoo email timeframe

1971 email released

Figure 1.3 Technological advances since the 1970s.

CHALLENGES

In our daily activities, we frequently work with strangers, even trusting them with our lives each time we drive our car, step on a plane, or get on a bus. We discuss important life issues such as finances, insurance, and medical assistance over the phone with people we don't know. People even fall in love over the internet. So, why is it so hard for us to adapt to working with people virtually? It's hardly surprising that some virtual teams are not successful. And even if they are successful, many virtual team members express a preference for working face-to-face. However, many face-to-face teams are not successful either. The reality is that virtual teams are rather new, and it may take some time to learn and adjust to the new environment and technology.

Common concerns about virtual teams include lack of facial and body language cues for validation, feeling isolated, and a demand for higher individual accountability because delays resulting from lack of preparedness are exacerbated and recovery takes longer (Shauna Wilson, "Forming Virtual Teams," *Quality Progress*, June 2003).

There is also the issue of whether people are visual or kinetic learners. Most consider themselves one or the other, leaving a small number who admit to being audio learners. The challenge of remote auditing for visual and kinetic learners is the disappearance of seeing other people in the flesh. Many of us rely on nonverbal communication for feedback and interpretation of what was said, and if what was said is true.

Common concerns about virtual teams include:[2]

- Missing nonverbal communication,
- Fear of isolation,
- Change in how we operate, and
- Demand for higher individual accountability because delays resulting from lack of preparedness are exacerbated and recovery takes longer.

Nonverbal communication can provide informal verification or affirmation. It's been said that we are so tied to nonverbal communication that about 93 percent of any communication depends on the nonverbal, not the verbal. Virtual environments limit this visual cue. However, nonverbal communication is subjective input and very judgmental. Some people point out that they are listened to better in a virtual environment than face-to-face because of their physical challenges or distractions. The virtual environment promotes auditing techniques that verify what is going on rather than interpreting nonverbal communication. Perceptions are important, but many times they fool us.

---

[2] Shauna Wilson, "Forming Virtual Teams," *Quality Progress*, June 2003.

> *Nonverbal communication
> is subjective.*

We use a variety of communication methods in our daily routine, such as sticky notes, action item lists, a wall calendar filled with events, and recognition awards collected from training. Though this communication is in passive message form, it is valued as a reminder of information, action, or recognition. The new virtual office can remove many of these traditional methods for organizing communication. In Figure 1.4, find examples of communication that will either change or disappear after transitioning from an active traditional office to a virtual one.

When teams go virtual, email or chat are heavily relied-on communication tools. On average, people receive one hundred emails daily.

Companies that use a virtual office software program can also use it as a backup for file storage and for messaging others in the virtual office.

Meeting frequencies can change, such as moving traditional daily face-to-face meetings to weekly virtual meetings because of time zone differences.

Virtual organizations must be prepared to make accommodations for the real differences between meeting online and face-to-face. Three important foundational components help virtual organizations be successful: cooperation, infrastructure, and technology.

COOPERATION, INFRASTRUCTURE, AND TECHNOLOGY

Working in a virtual team is a lot like driving a car. It requires the cooperation of others, sound infrastructure, and reliable technology.

Cooperation

We trust that other drivers on the roads we share are trained, healthy enough to drive, and are not under the influence of substances, texting, or incapable of operating their vehicles. With a common goal that all

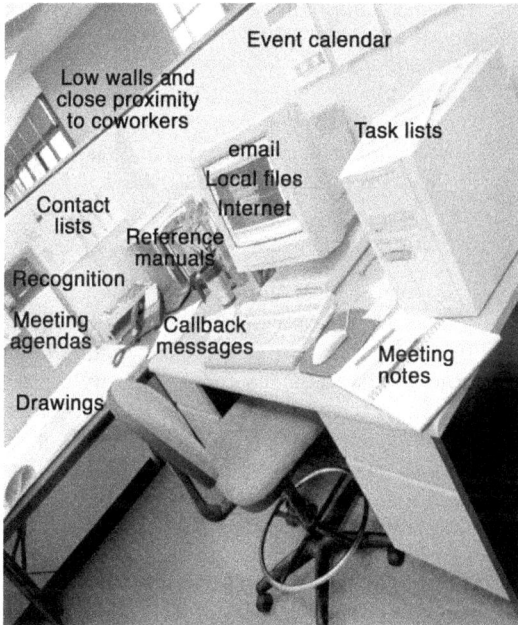

Figure 1.4 Communications may change or disappear when transitioning to a virtual office.

successfully reach our destinations, we cooperate by letting others into our lane, keeping a safe distance behind the person in front of us, and obeying the laws of the road. Similarly, virtual team members should forgive misunderstandings, proofread emails for how the message could be misinterpreted, or call someone to talk about process issues when needed. We cooperate and communicate with one another to reach common business goals and objectives. In this book, you will learn communication theories and solutions that help improve our communication while remote auditing.

> *Virtual team members forgive misunderstandings, proofread emails for potential misinterpretation, and are willing to call someone when one-on-one voice communication is needed to resolve process issues.*

Infrastructure

The road system's standards and regulations save lives. If we stop and think about all the cars and trucks that travel our interstate highways, freeways, and other roads, it is the infrastructure, including speed limits, signage, training, and certification—supported by enforcement—that ensures this road system continues to move. A virtual organization should pay no less attention to infrastructure requirements like fast, secure broadband to ensure its success. A virtual organization's internal audit process using remote auditing methods is an excellent means to gauge performance.

We will define audit models and processes that replicate the working environment, meet an immediate need, and support the audit process.

Technology

Automotive systems can now warn you if there is something in front of, behind, or beside you; reader boards can provide an estimated time of arrival in bumper-to-bumper traffic; and the list of technology advances rolls on. Our new road map as auditors is to keep up with the technological changes, both on the roads and from our home offices. A challenge for virtual organizations is selecting technology that enhances communication and can interrelate to the relationship level of remote members globally.

In this book, you will learn how to use collaborative technology to conduct online interviews and use cameras for visual assistance.

CHAPTER REVIEW

- Telecommuters can experience the following issues when working in a virtual organization:

  A. Delays in decision-making

  B. Feelings of isolation

  C. Difficulty with building trust

  D. Challenge of validating without verbal communication

- A remote audit uses the same audit process:

  A. Prepare

  B. Perform

  C. Report

  D. Follow-up

- The changes in the world that have significantly altered the way we conduct audits include economics, global marketplace, and technology.

- An advantage of remote auditing is that it replicates the new global work environment; thus, the auditor is experiencing a similar virtual work environment to that of the auditee. Other advantages of remote auditing include:

  A. Saves up to one-half of a year's auditing time

  B. Provides real-time auditing, especially for supplier audits

  C. Minimizes delay due to travel

  D. Fewer cancelled audits due to unforeseen incidents (i.e., extreme weather or political conditions)

  E. Reduces the time to write audit reports because auditors are recording as they audit

- Among the reasons for negativity about virtual teams are:

  A. Missing nonverbal communication

  B. Level of technical skill needed

  C. Change in how we operate

  D. Demand for higher individual accountability because delays resulting from lack of preparedness are exacerbated, and recovery takes longer

- The disappearance of the office impacts passive communication. Passive communication includes wall calendars that note events, action item lists, and contact lists; all provide reminder information to the office personnel.

- People management change from hours in the office to meeting deliverables has prompted online tools that monitor work status.

- A lot like the transportation system, virtual teams need cooperation from the workforce, a solid infrastructure, and current technology to be successful.

Chapter 2

Virtual Communications: Theory and Practice

Understand communication methods used in the virtual setting and how to use these methods to build trust, interview, and ensure validation.

1. Learn applicable communication models and theories for auditing online.

2. Review important trust-building communication processes.

3. Apply the appropriate technology to the correct type of communication.

COMMUNICATION MODELS AND THEORIES

Communication Challenges

A problem occurs when people use traditional meeting methods in the online meeting environment. Using the internet can make remote communication much different and more difficult to understand when compared to members occupying the same room. Even knowing when to start talking can be an issue for remote meetings or interviews.

When communicating over the internet, interference noise can stem from the technology used (bandwidth issues, compatible software, and hardware) and limited or missing nonverbal communication. This inability to observe natural reactions may cause uncertainty and anxiety in some people. For others, the virtual environment may result in less tension (no one watching them), and they may share things they would not if the meeting were face-to-face.

The communication process begins with a sender initiating a message to a receiver. Then, receivers decode the purpose of the message, try to understand the message, separate the noise that may exist, and then interpret the message. Receivers will provide feedback to the source, asking for clarification of the message, clarification of their interpretation, or stating technical issues (see Figure 2.1). Having said this, we know that not all messages are the same. Messages may allow limited feedback, or perhaps assumed or one-way communication. Some only require acknowledgment, while others require verification that the message was received and understood.

To communicate, the encoder starts transmitting information and sends the message to the decoder. Clarifying questions are used as feedback to understand the message better.

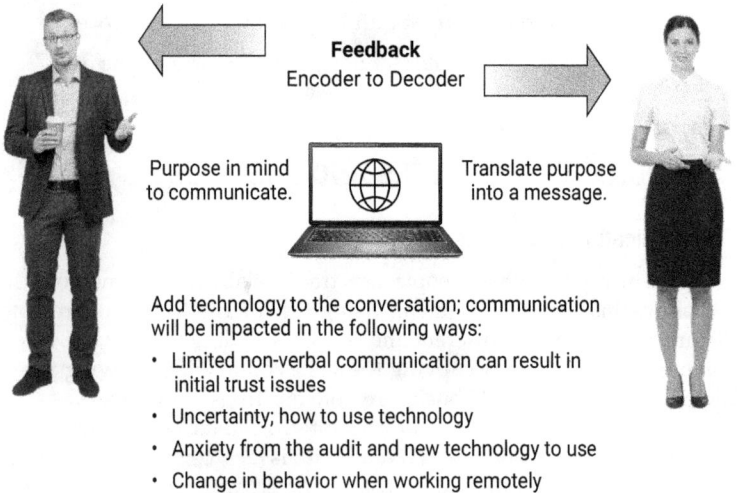

Feedback
Encoder to Decoder

Purpose in mind
to communicate.

Translate purpose
into a message.

Add technology to the conversation; communication
will be impacted in the following ways:

- Limited non-verbal communication can result in
  initial trust issues
- Uncertainty; how to use technology
- Anxiety from the audit and new technology to use
- Change in behavior when working remotely

Figure 2.1 The communication process.
Source: Modified from the Shannon Weaver model.

Two communication theories, uncertainty reduction theory (URT) and self-monitoring behavior theory, provide insight into the challenges of working online.

Uncertainty Reduction Theory (URT)

When people meet online, they lose the ability to communicate nonverbally. Nonverbal communication is important as we often *look for signs of agreement or approval* in the faces or body language of others. When we lose the ability to communicate nonverbally, an uneasy atmosphere is created, which increases personal uncertainty.

> **Uncertainty in a virtual environment**
>
> *Nonverbal communication is limited or nonexistent.*
>
> *Computer anxiety—technical difficulties*
>
> *Keyboarding skills—inability to keep up or hitting the wrong keys*
>
> *Less immediate anxiety—could be delayed (worry about how an email is perceived)*

To reduce uncertainty, many people find comfort using the camera to see who they are conversing with. This may help, but because of internet bandwidth issues, video may last for a limited time or not at all. To resolve uncertainty, talking and finding commonality can help to calm nerves. Anxiety will occur most frequently over the use of technology.

As auditors, we need to be aware of the following negative reactions most felt by the auditee during heightened anxiety levels:

- People *feeling technically incompetent* may express frustration.

- Fear of *negative behavioral consequences* causes distrust, exploitation, and manipulation.

- Fear of negative *evaluations by strangers* creates feelings of being rejected, ridiculed, or disapproved.

VIRTUAL COMMUNICATIONS

Self-monitoring Behavior

The second theory we want to discuss is self-monitoring behavior. Self-monitoring behavior *maintains that some people are sensitive to how they are perceived by others, while others are not.* Individuals with high self-monitoring behavior pay more attention to how they are viewed by others, even in online environments. These people are quicker to ask questions and *provide validation*, which helps them adapt to meeting online more easily compared with the low self-monitoring individuals.

Folks with low self-monitoring behavior often appear semi-absent from meetings because of multitasking (taking a phone call, checking email, returning late, or doing other work). This behavior may be perceived as hostile because they are not as engaged and involved in the meeting. In this environment, miscommunication occurs, a false sense of security develops, and *distrust is validated.*

Silence During Communications

Silence is often *caused by multitasking participants* reading their email, texting others, or surfing the web. *Silence could stem from not knowing when to talk.*

> ### Low-monitoring behavior
> *More likely to blurt out something that would not be said in person.*
>
> *Multitasking!*
> - *Texting*
> - *Sending emails*
> - *Allowing for disruptions*
>
> *They may be late to rejoin after breaks.*
>
> *Behavior may seem hostile and disengaged.*

Silence also results from the auditees or auditors not having the correct skill set for communicating over the Internet. Some silence in an online meeting is normal, but too much should be avoided. *Keeping communication going is more important in an online environment* than watching for nonverbal communication. In a way, one may consider nonverbal communication in a face-to-face meeting as a stimulus, such as people moving, getting up and leaving the room, changing positions, and so on. Hence, techniques need to be used to keep online meeting participants engaged.

BUILDING TRUST IN REMOTE AUDITS: THREE-STAGE DEVELOPMENT

When we are introduced to someone new, we typically step through these *three stages to build trust* (see Figure 2.2). Initially, if available, we ask people we know about this person or perhaps look them up on LinkedIn.

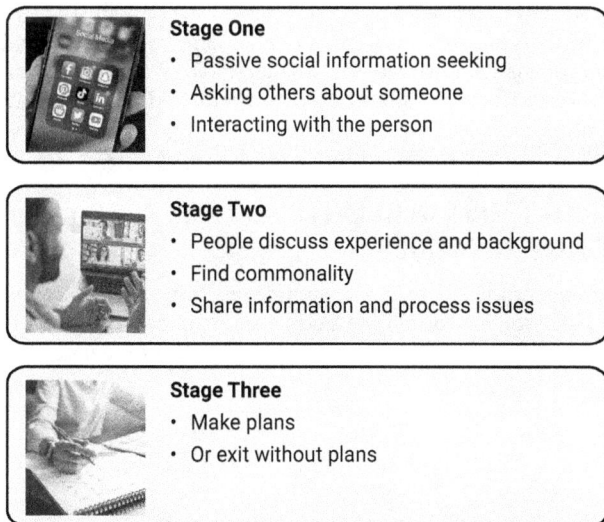

Stage One
- Passive social information seeking
- Asking others about someone
- Interacting with the person

Stage Two
- People discuss experience and background
- Find commonality
- Share information and process issues

Stage Three
- Make plans
- Or exit without plans

Figure 2.2 The three stages to building trust.

When we meet, we begin a conversation looking for some type of commonality— perhaps where someone is from, what companies they have worked for, or where they went to school. Auditors can spend the first five to ten minutes of an opening meeting looking for commonality and building trust with the auditee. Perhaps talk about the weather or the auditee's place of residence. Who knows—it may be a place you want to visit on a vacation. Small talk needs to be gender-, race-, and politically neutral. Avoid geographical bashing and negativity.

> *Typically, when we first meet someone,*
> *we look for some type of commonality.*

Next, we make plans to accomplish work together. Now is a suitable time for you, as the auditor, to clearly define the audit process and discuss the audit plan. Trust can be built by setting clear audit processes, so the auditees understand what is expected from them, and by asking a few questions to learn more about the person you are auditing. Ensure that everyone is comfortable with the technology and understands the audit process. The audit time should be conducted within the auditee's time zone.

BUILDING TRUST IN REMOTE AUDITS: MEETING PROTOCOLS

By *outlining expectations in the opening meeting,* auditors can define proper behavior for the online audit. Auditors can create an inviting environment by calming nerves and setting behavioral expectations.

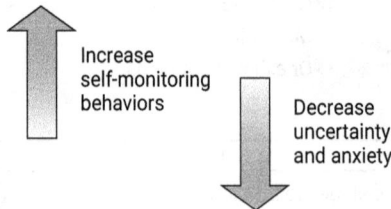

Increase
self-monitoring
behaviors

Decrease
uncertainty
and anxiety

To calm nerves and decrease anxiety, consider the following:

- Conduct technical testing prior to the audit
- Send an audit plan well in advance
- Spend a little time at the beginning talking to the auditee
- Create an open environment for questions

Set Expectations for Proper Behavior

Determine the meeting rules before virtual meetings. Example meeting rules include:

- Take attendance for opening and closing meetings.
- Ask auditees to inform others that an audit is in progress and to hold their calls or interruptions until after the audit.
- Ask for cell phones to be turned off during the audit. No texting during the audit; unless necessary for data collection.
- Ask auditees to respect start times and arrive at the audit sessions on time. This is very important because you cannot go down the hallway or into the office next door to remind someone that the meeting has started. You can lose control of a virtual meeting if scheduled times are not honored.
- Request a quiet area or room for the virtual meeting. If there is a lot of background noise (like equipment, people talking, dogs barking, and even babies crying can occur in home-based offices), it may be disruptive.

All *auditors* have sat in meetings where there is silence. Not being in a room with the auditee and *knowing when to talk is sometimes challenging.* Here are a few tips to keep participants engaged:

- *Provide a clear meeting structure or agenda* so others understand what the next steps will be.
- Tactfully let the auditee know when it is time for them to speak (for example, say, "Your turn now. It's time for feedback. What do you think? How do you see it?"). If you need a particular person to speak, you need to use their name or

other identification since there are no nonverbal clues for others to identify.

- Use the *round-robin technique to ensure that all auditees have been given a chance to talk.* The collaborative program will list the people who have joined the meeting— you can arrange the round-robin starting at the top of the list.

> *Prior to starting the meeting,*
> *set the meeting rules.*

BUILDING TRUST IN REMOTE AUDITS: STORYTELLING

If anxiety levels are heightened in your audit (especially when technological issues exist), interview questioning may escalate the anxiety. Since most people are visual learners, and everyone enjoys stories, encourage participants to show you examples or draw what they may be trying to explain. *Storytelling* is another technique to help folks feel more comfortable in a virtual environment. David Cooperrider's appreciative inquiry (AI) method of questioning can be a useful tool for generating discussions and gathering data at the same time.

The *AI process* (see Figure 2.3) consists of five main steps:

1. *Definition.* Establishing the focus and scope of the inquiry.

2. *Discovery.* Eliciting stories of the system at its best.

3. *Dreaming.* Collecting the wisdom and imagining the future.

4. *Designing.* Building bridges to the future based on the best of the past and the present.

5. *Destiny.* Making it happen.

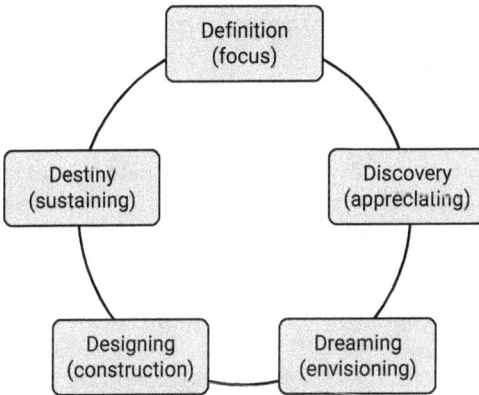

Figure 2.3 Steps in the appreciative inquiry method.

Here is an example of applying a line of questioning for auditing using AI techniques:

- Describe an event that best illustrates customer satisfaction.
- Ask how the process helped with that success.
- Ask participants what they value most about being a member of this team.
- Ask them what they would change if they were to do something over again.
- Ask them how they would implement those changes.

Eliciting stories of the system is the discovery step for the AI process.

BUILDING TRUST IN REMOTE AUDITS: THE DIRECT METHOD

For those who work better with the direct approach, *begin with an open-ended question,* such as "Could you please explain your process to me?" After the auditee provides a process overview, continue. This time, *use clarifying questions* as you step through the process. Clarification questions are more direct and are at times the same question asked differently. Listen for continuity and repeatability in the answers and the evidence. Auditees that like structure or organization may prefer a more direct line of questioning as the auditing method.

This line of questioning flows in a sequential format that, in the end, tells a story of foundational structures built to support processes (see Figure 2.4). Although this sequential format is different from the storytelling method, both will achieve the same result if the method matches the personality of the auditee. It is important to realize these differences to capture the real process while gaining trust from the auditee when auditing online.

If you find remote auditing challenging, I recommend you practice using the technology until you feel comfortable with it. Practice drawing flowcharts, mind maps, and affinity diagrams on an online whiteboard. This skill will help you in online meeting facilitation so you can help an auditee draw or better communicate what they are trying to explain to you.

TYPES OF COMMUNICATION

There are two dominant communication methods that drive how people communicate with one another either face to face or online (see Figure 2.5). The most familiar is the *synchronous* method because we use it in our face-to-face conversations. We obtain instant feedback, either through words or nonverbal communication. We use these methods to understand meaning and validate truth.

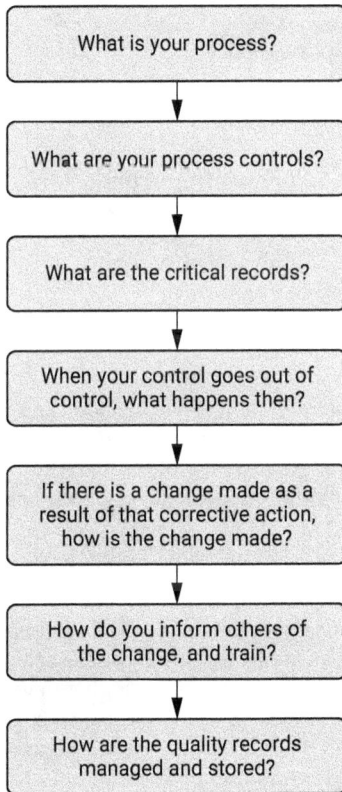

Figure 2.4 Flowchart of sequential questioning format.

We use *asynchronous* communication when writing notes and letters to one another. In some ways, it can be more difficult to understand meaning and validate truth using this method because of the absence of instant feedback for clarification or the interpretation of nonverbal communication. When we audit, however, asynchronous communication can be considered a record of the communication or action taken, like an email message, an entry on a form, or a document.

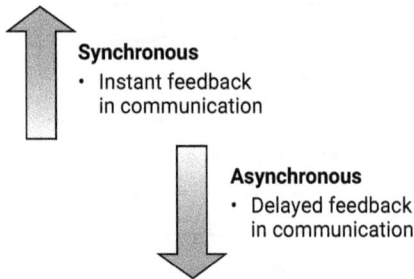

Synchronous
• Instant feedback
  in communication

Asynchronous
• Delayed feedback
  in communication

Figure 2.5 Two dominant communication methods.

As a culture, we interchange synchronous and asynchronous communication continuously throughout the day as we work and live. We form favorites, being exclusively face-to-face for important matters. Others may send a text message, email, or phone call.

Communication Types and Applicable Technology

The type of technology matters and is an important factor to consider when auditing because people will naturally fall back on their favorite technology in crisis, or when auditing issues such as timely corrective actions that need to be escalated to higher levels of management. Messages could be missed if the same technology is not used, and then communication could be bypassed. Figure 2.6 identifies the communication medium used in each communication type.

Determining when teams use the phone, discussion boards, and email to deliver information helps communication in remote teams. *Communication matrices* address the communication path one can take to inform others when issues arise. When an issue needs to be escalated, *the matrix* can be used to define contacts and the technology to use with the associated time zone. Escalation process definition reduces miscommunication in crisis. Overcommunication caused by duplicating the same message using voice mail, email, and fax is counterproductive.

Asynchronous (delayed)	Synchronous (instant)
Voice mail	Instant message (IM)
email	Telephone
Internet offices	VoIP
Websites (blogs)	Collaborative meeting room

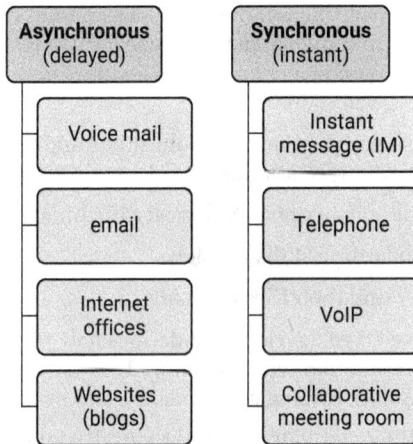

Figure 2.6 Communication media and types.

CHAPTER REVIEW

- One area that impacts communication initially is the loss of nonverbal communication. For some, it can consume their attention.

- Asking more questions tends to clarify unknown aspects, thereby increasing trust levels.

- Self-monitoring behaviors without employing face-saving devices can result in people behaving differently online than in person. Setting expectations and rules, while providing information about processes to lower anxiety, will help keep behaviors in line.

- Low self-monitoring behaviors may seem hostile and make the participant seem disengaged from the task at hand.

- Uncertainty reduction theory (URT) assumes everyone enters an unknown area with some level of fear or anxiety. Some common fears or anxieties auditees have in remote audits include:

    A. Missing nonverbal communication; auditees cannot see the auditor for assurance when answering questions.

    B. Computer anxiety—technical difficulties.

    C. Keyboarding skills are slow.

    D. Can't find their files or emails.

- The three-stage development defines how people learn about one another when meeting. These stages include passive investigation, finding commonality, and making plans.

- Storytelling methods are ideal for understanding processes; the five appreciative inquiry (AI) steps that can be used for questions are definition, discovery, dreaming, designing, and destiny.

- The direct auditing method leans more toward sequential thinking and is more linear.

- Auditees who like structure or organization may prefer a more direct line of questioning as the auditing method. Persons may be abstract sequential and concrete sequential.

- Auditors need to select the asynchronous and synchronous communication that should be used as part of the audit process so that messaging and communication are consistent.

Chapter 3

Technology—Interview and Record Review

Learn how to use collaborative technology to conduct online interviews and record reviews:

1. Introduce collaborative technology for interviewing—desktop sharing, whiteboard technology, and keyboard sharing.

2. Learn how to use the technology in an audit.

3. Define online meeting protocols and common issues when working with technology.

4. Introduce asynchronous technology often seen in record reviews—virtual offices, shared drives, databases, email.

5. End-of-audit housekeeping—document and information security, verify follow-up actions.

TECHNOLOGY USED FOR SYNCHRONOUS (INSTANT) COMMUNICATION

Figure 3.1 shows an example of a menu bar that to use to *manage a collaborative meeting room*. Once you understand what the icons represent, using this tool will become easier. Think of it as a meeting room. When you log in, it is like turning on the lights. Collaborative rooms provide auditors with more options for facilitating a meeting than a face-to-face room, such as being able to mute the microphone or do an instant message for follow-up records needed.

Collaborative meeting rooms provide a very real experience and an efficient means for conducting remote audits. A menu display will appear on your screen, letting you know when your online meeting is live. Icons are similar across the collaborative programs. The following are a few common icons needed when auditing remotely.

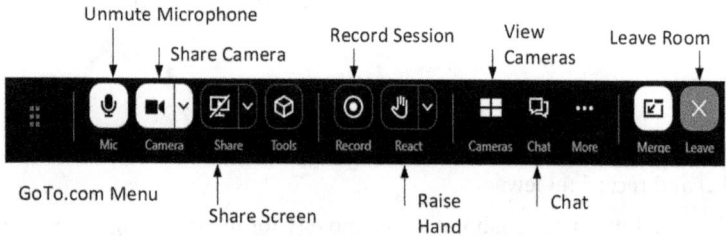

Figure 3.1 Example of virtual meeting room menu bar.
Source: GoToMeeting.

Here are a few of the collaborative program functions and uses detailed:

- Screen share allows for multiple documents to be shared. Program share will only present a single document requiring you to stop sharing before showing another document.

- Change presenters (turn over control to the auditee, who can use the software tools such as the whiteboard, or display records or documents).

- Highlighting pens and markers can be used for notation in shared documents or diagrams.

- Keyboard share is a tool for giving access to someone else to search for files or keyboard notes.

- Chat function is a handy tool for private or public notes. Public notes can be used as a follow-up list of action items.

- Camera sharing allows everyone to see one another, observe a process, or illustrate the functions of a product.

- Computer mic and speakers eliminate the need to dial into the meeting conference. However, better quality is noted if all participants wear a computer headset and microphone.

- Phone—Use the provided collaborative phone number assigned to the meeting room.

- Screenshot—electronic capture of what is shared on the screen in a graphic file. Permission should be requested before taking a picture of the document shared.

- Viewers zoom in and out will enlarge the viewing area like print so now it's readable. Zoom-out features will allow for an expanded viewing area.

- Whiteboard tools are useful for brainstorming, drawing, or highlighting on floor diagrams and documents.

Equipment Needed

- USB computer headset (includes microphone and audio)

- Mobile device: smart phone, tablet

- Laptop (aged computer equipment will only slow the connection)

- Desktop with microphone, camera, and speakers added

- For best performance and security protections, ensure all platform and security downloads are current on all devices.

Collaborative Programs

There are many collaborative program options to choose from (GoTo, Teams, Zoom, Google Workspace); do research for the short-listed tools; but also consider the following criteria:

- Cybersecurity controls—Is the collaborative program encrypted? Is there an option to lock the meeting room or set passwords to join? When linking mobile devices to the collaborative program, does it affect the viewing quality of the remote location?

- Consider using a *virtual private network* (VPN) protected network when using Wi-Fi.

Most companies provide similar functions (but not all the tools on the shortlist) and free trial periods. When choosing a collaborative room, look for ease of use, and ensure that the functions offered meet your auditing needs and equipment requirements, including internet speed.

FEATURES OFFERED IN A COLLABORATIVE MEETING ROOM

Table 3.1 Summary of collaborative meeting tools for use in remote auditing.

Features	Function(s)	Remote auditing uses
Screen sharing (includes application sharing)	View another's computer synchronously	Review records, procedures, and document storage on another computer
Instantly change presenters	Ability to change to another presenter's PC for screen sharing	Switch from one PC to another to follow audit trails
Share keyboard and mouse control	Let someone else drive!	Highlighting areas on documents
One-click recording	Record the meeting: Permission to record should be considered	Ideal for keeping closing meeting records as an audio or video file
Whiteboard drawing tools	Ability to highlight or draw arrows	Use it to clarify and describe what you are talking about by drawing it
Audio conferencing (via telephone and computer)	Audio using either VoIP or phone *Note:* Headsets help with overall sound quality	Others (such as management) can attend the audit or remotely observe

HOW TO USE COLLABORATION ROOM FUNCTIONS DURING A REMOTE AUDIT

Opening Meeting

Set up the virtual meeting room by ensuring the microphone and camera are on (to see other attendees and yourself) to use for introductions (see Figure 3.2). Also, turn on the share screen option to display the opening meeting sign-in sheet. It's a good idea to obtain the attendee names and their titles before the opening meeting.

As people come into the collaborative room, greet them as you would in a face-to-face situation. If sound and cameras are not working, ensure the mic and speakers are correctly selected in the collaborative program (see Figure 3.2).

When using Share Screen, anything that appears on that screen will be visible to everyone. Share the completed attendance list; an easy roll call can begin the opening meeting. If their name does not appear on the attendee list, it's an easy update. There are other methods to take attendance, but using the sign-in sheet on the screen immediately engages attendees in the virtual meeting environment and equipment aspect. Once introductions are completed, the auditor should cover the audit plan, make necessary adjustments, and state audit guidelines. Figure 3.3 illustrates the process highlights.

Using Share Screen

After the opening meeting, attendees who are not needed for the next agenda item can log out. If they need to come back to the meeting, then they will log back in using the same link used for the opening meeting.

You will need to *provide presentation permissions to the auditee* by selecting the person in the contact window and clicking the presenter icon. The auditee will need to agree by clicking an affirmative answer. You are now staged for conducting your remote audit (see Figure 3.4).

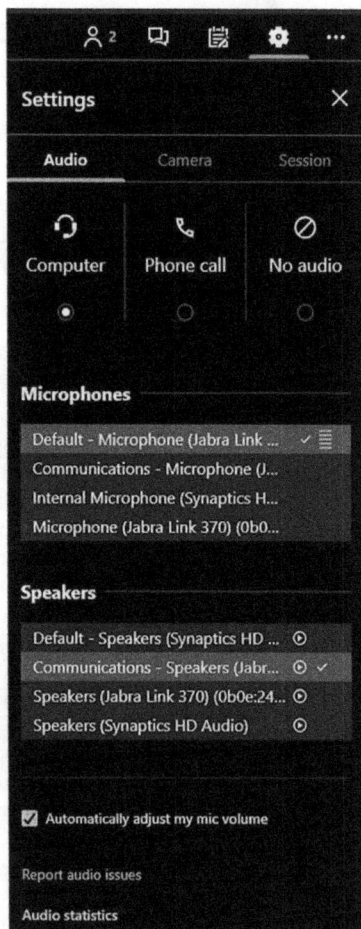

Use the Utility icon to verify the audio and camera selected. If there is a problem with the sound, ensure the correct microphone and speakers are selected. If computer audio is not functioning, select Phone Call for the meeting call-in number.

Figure 3.2   Virtual meeting room audio and camera settings.
Source: GoToMeeting.

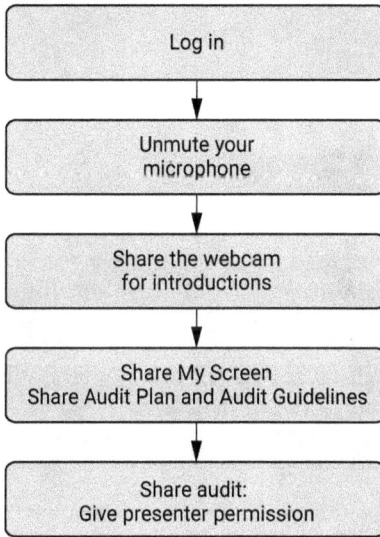

Figure 3.3 Virtual opening meeting process.

When selecting the share screen option, two initial options are given: share screen or application. Sharing the application applies to one document at a time. Sharing the screen allows the auditee to move to several documents throughout the interview without having to unshare and reshare an application. There may be an option to choose the monitor for sharing, too.

At the end of the session, change the presenter back to your name until the next auditee arrives. Being able to manipulate the software (whichever you are using) is an important skill for conducting remote audits. Refer to the following flow diagram of the remote audit process (see Figure 3.5).

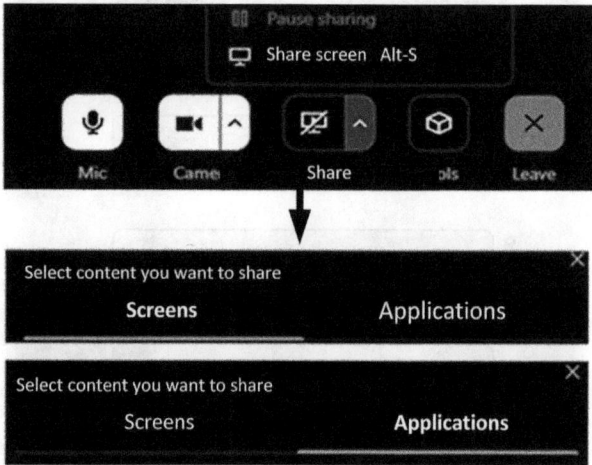

Figure 3.4 Sharing screen or applications.
Source: GoToMeeting.

> *At the completion of the opening meeting, people who
> are not needed for the next agenda item should log out.*

One more useful tool to understand is the *whiteboard*. Whiteboards
are highly interactive, allowing attendees to draw what they are talking
about. Not all whiteboards are as visible as the display in Figure 3.6, but
if you locate drawing tools and pen colors (as shown), you and others
then could draw using the keyboard share function. Not only can you
paste clip art and flowchart symbols onto the whiteboard, but when
you select A at the top left of the drawing tools (see Figure 3.6), you can
keyboard text, too. To do successful remote audits, you need to practice
using the software.

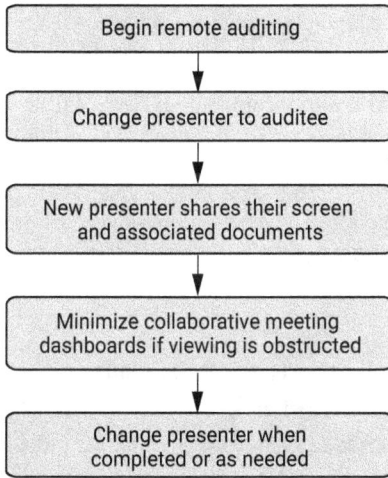

Figure 3.5 Remote auditing process.

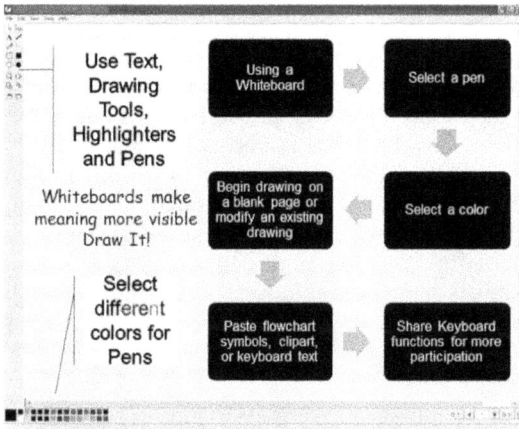

Figure 3.6 Virtual whiteboard example.

USING EMAIL FOR RECORDS

It is best not to use email for records because of traceability issues. It is often not clear who is either copied on the email (that is, blind copies) or to whom the message is eventually forwarded. Another problem is that when specific personnel is assigned approval duties and they use email, then there is an access issue, especially if that person is gone and no one else can get into their email because of passwords. Last, but by no means least, the sheer volume of emails received daily is impossible to manage, and people can easily miss a critical message.

Issues with email

- *Thousands of emails*
- *Hundreds of email folders holding thousands of emails*
- *Forward/reply to all*
- *Forward/reply to one, not all*

When reviewing a person's email filing system, you may find very good record storage management. However, areas that cause problems include the distribution tracking of some emails and allowing others to access email records. These issues can be resolved by defining a minimum required distribution list and providing greater access to specific email addresses. For instance, many companies assign an email address such as info@email.com, giving a specific number of people the proper authority to access the email.

Table 3.2 lists some pros and cons of using email transmittals as a distribution record.

VIRTUAL OFFICES (COMPUTER-BASED OFFICES)

Another distribution method to use instead of email is centralized communication on either a website or portal. Often called *virtual offices,* these programs have attributes similar to those of brick-and-mortar offices. Virtual offices are the answer to the lost communication posted in a traditional face-to-face office and the massive email issue.

Virtual offices are powerful tools for centralizing communication within discussion boards and managing files with check-in and check-out controls. They include databases that can be used to track issues. Everyone with access has visibility to what is shared in an internet office.

Table 3.2 Pros and cons of using email for record keeping.

Pros	Cons
Used most of the time	Not traceable because emails branch off via forwards and replies; the sender does not have visibility to all the receivers
Convenient	Limited access—if processes such as an approval process using MS Outlook are used, only the initiator has access to all the approvals
Fast if emails are well-managed	100 daily emails are difficult to manage
	Lost email

Another great feature includes status bolding—letting you know instantly when something has changed—and/or the capability to push change notifications to a distribution list. These offices have keys and allow only those with access to enter or make changes. Figure 3.7 displays a screenshot of a virtual office portal.

The nemesis of virtual offices is email. It takes discipline to upload documents and use threaded bulletin boards. It is much easier to email an attachment.

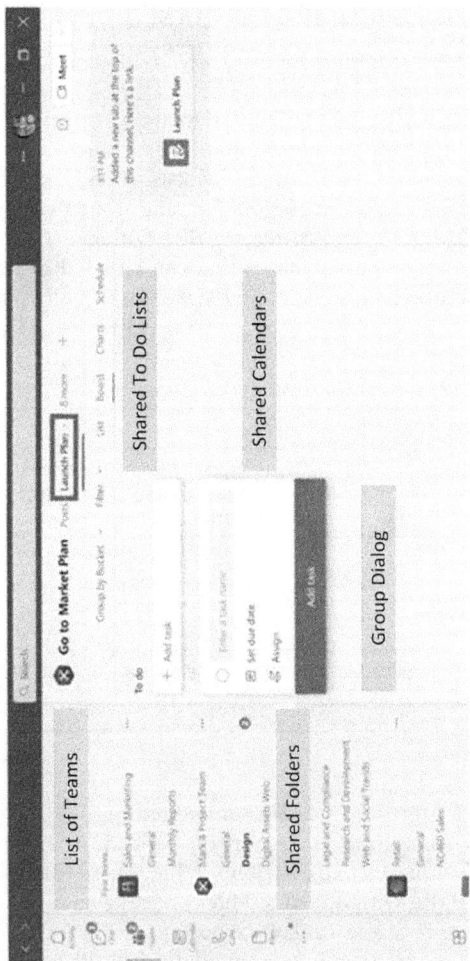

Figure 3.7 Virtual office example (Microsoft Teams).

ASYNCHRONOUS COMMUNICATION TOOLS

Table 3.3 provides the common features and uses found in online offices, such as Microsoft SharePoint, Teams, QuickBase, JobBOSS, and SalesForce. The third column suggests some of the auditing and management system tips for these features.

Table 3.3 Common features and uses of online offices.

Features	Function(s)	Auditing and MS uses
Folders	Stores shared files. Properties include check-in and check-out functions and revision control	Uniformity—sequential (as stated, that is, monthly, quarterly), document access control privileges, information folders or topics
Discussion boards	Centralized location for discussing specific topics; responses are threaded	Traceability of discussion and timeliness
Calendars	Display important dates on the virtual office calendar	Internal communication meeting or event dates, and verification of dates
Lists	The database used for suggestion boxes, meeting agenda items, to-do lists, action item lists, and so on	Management review agendas and internal communication action items
Surveys	To collect data for decision-making, understanding team opinions, or as a voting technique	Management responsibility; internal communication feedback tool
Push technology	Request notification for changes in the office. This notification can be delivered via email, pager, or cell phone	Change control notification; internal communication

PREPARING THE ATTENDEES BEFORE THE AUDIT

Here are a few helpful protocols to define when working in a collaborative meeting room. Because audits can take place in the auditee's home, respect privacy by advising everyone to:

- Mute audio when background noise increases (barking dog or crying baby).

- During breaks, mute your microphone and turn off your camera; in fact, only use these tools when you are in front of your computer ready to participate.

- Sync time before breaks to ensure that all return on time. Having everyone's phone number on hand is always a good backup.

- Dress accordingly, not eating or smoking during the audit.

- Ensure the background setting is business professional.

- Ensure privacy within the local area—isolating yourself from others in your household.

END-OF-AUDIT HOUSEKEEPING

At the end of the audit, it is the lead auditor's responsibility to safeguard all documents. For a remote audit that uses computer technology and the internet, there are safeguard challenges. The lead auditor should:

- If available, use a secure internet site like Dropbox to transfer files, designating access accordingly.

- Remove from electronic devices or otherwise secure supporting documents and materials belonging to the auditee according to the agreed-on arrangements and applicable statutes and regulations.

- Securely store or destroy any created reports, computer files, graphics, videos, and so on, according to the arrangements and applicable statutes and regulations.

- Confirm confidentiality by reminding all parties—including other auditors and language, technical, and clerical support staff—that data collected, conclusions, and all other aspects of the audit are confidential, and that any disclosure of information could result in undesired consequences.

An auditor's individual PC may have limited security and may be subject to hacking. Retained audit documents can be stored in a designated Dropbox (cloud), or the files could be transferred to the auditing organization's secure site, if available. Table 3.4 summarizes concerns about electronic report distribution.

Table 3.4 Summary of report issues and concerns.

Issue	Concern
Control of distribution	Once in an electronic format and put on the network, reports can be sent to anyone.
Report errors	Errors in grammar, spelling, and word usage can increase unless reports are proofread.
Not concise	Reports can become lengthy when boilerplate information is cut and pasted into reports.
Tampering with report content	Unless content is secured (via password or PDF security options) report content can be modified and information integrity can be lost.
Recalling reports	Storage medium may become obsolete, or systems become corrupt, and make the reports inaccessible or too expensive to access. For example, tape storage or server failure due to a virus or mechanical failure.
Removing (destroying) reports	On the one hand, it is easy to lose information in electronic format, yet if you want to remove certain information, it may be nearly impossible to destroy. It may always be on someone's system.

FOLLOW-UP ACTIONS TAKEN

When conducting internal audits and some external audits, follow-up activities may be accomplished remotely. Remote audits verification allows for timely follow-up in an economical manner. An organization may decide to use remote audits techniques to follow up all auditee actions. The remote audit could also be used in conjunction with the auditee providing the auditor with periodic reports and records containing evidence that the corrective action plan has been implemented and is effective. In this way, the auditor may be able to verify that the action plan is effective without the added expense of another on-site audit. The auditor still has the option of verifying the long-term effectiveness of these activities during subsequent visits.

Network Security

While auditing, it is the auditor's responsibility to ensure that information is secure over the Internet:

- Use the auditee company's protected Wi-Fi for mobile devices like tablets or phones (camera transmission).

- Working from home, a hardwired internet connection is safer than a Wi-Fi connection (password protected).

- Use a virtual private network (VPN) for additional security in a public domain.

- Keep computer virus protection updated at all times.

- Use proprietary collaborative applications like GoToMeeting, Google Workspace, Zoom, or Microsoft Teams.

CHAPTER REVIEW

- For best performance when you VoIP, use a USB headset that includes a microphone and audio functions.

- Whiteboards are the collaborative technology function that provides the highest level of interaction between remote attendees.

- Using synchronous tools to interview the auditee while reviewing asynchronous records is the best way to conduct remote audits.

- Obtain permission for screenshots during the audit and for any meeting recordings needed.

- The screen share function has earned the reputation, "It's better than being there in person" because the best view of the auditee's records is from your own computer monitor.

- Internet office functions store passive information, as well as highlight changes to information and manage changes.

Chapter 4

Technology—Visual Tools

L earn how to use collaborative technology and cameras as visual tools:

1. Introduce collaborative technology
   - Cameras
   - Smartphones
2. Learn how to prepare for a visual review
3. Introduce how to set up viewing locations and instruct the remote person
4. Best-practice tips in preparing for a visual review

USE OF CAMERAS IN REMOTE AUDITING

The old adage "A picture is worth a thousand words" comes to mind when viewing processes using cameras at remote locations. Cellphone cameras are used to take photos of checks to deposit into the bank via a text message. Surgeries are performed remotely using cameras, allowing patient access to leading physicians. *Cameras and the internet, when used in harmony, provide real-time options* for many different uses, including auditing. This section will provide checklists and insights to help you prepare for remote visual observations. You will learn about:

1. The cameras to use as a viewing tool at remote locations

2. The different technologies and interconnections that are required to successfully broadcast a live video stream over the internet from a remote location

3. How defining viewing angles and "shooting" locations at the remote site is instrumental to the quality of the visual review

Three process elements need to connect for successful remote viewing (see Figure 4.1). Though *all three components are necessary to view remotely,* one component is a deal breaker if not available: Wi-Fi, and secure broadband Wi-Fi strength needs to be a few bars at its maximum capability to support both audio and video. If Wi-Fi is weak, add a booster in the area to ensure clear viewing and audio.

An outward-facing camera for viewing the operation or activities provides both the auditee and auditor the ability to see the same view simultaneously. Using Wi-Fi and a securely connected camera feeds the video stream from the camera to the internet. Then, streaming via a collaborative program (e.g., GoToMeeting, or FaceTime, Teams) allows the auditee and auditor to communicate and review remote locations.

```
┌─────────────────┐     ┌─────────────────┐     ┌─────────────────┐
│    Cameras      │     │ Secure internet │     │                 │
│ • Outward view  │ ──▶ │  connection to  │ ──▶ │  Collaborative  │
│ • Inward view   │     │  stream video   │     │    program      │
└─────────────────┘     └─────────────────┘     └─────────────────┘
```

Figure 4.1 Three components necessary for successful remote viewing.

> *It important to have an outward facing camera for touring remote locations. Everyone sees the same thing!*

CAMERAS

The best cell phone camera is an Apple iPhone. Ask the auditee to find one for the audit ahead of the audit. Alternatives will be a laptop or another type of cell phone with a lesser-quality camera. If a laptop is chosen here are the items to be handled:

1. The auditee does not have a good view of where they are pointing the camera and what is streaming to the auditor because the laptop is turned away from the auditee handling the laptop.

2. It is difficult to hear the auditor speak with the laptop turned away due to the inward laptop camera.

3. It requires the auditor to continually voice directional instructions to prevent ceiling and floor shots and moving too fast.

4. Another challenge to using a laptop has to do with the physical size and weight causing fatigue.

Tablets will work as well if they meet the same requirements as a cell phone (Wi-Fi access and the ability to log in to a collaborative room. We know technology changes often, so keep on top of the changes because there may be an easier way to do something with the equipment you already own.

Many people believe an Apple iPhone, iPad, or MS Tablet to be the best tool for providing a clear view and ease of use at remote locations. To stream video through FaceTime, you will need Wi-Fi access for your Apple product. To connect, simply call the remote person using FaceTime on your Apple product using their phone number.

Of all the current smartphones, an iPhone is an excellent tool to use for floor tours and visual process reviews. The clarity is very impressive. Other devices may present a blurry representation, making it difficult to read items on the production floor. This will cause an audit delay as hard-copy records will need to be scanned and shared. It is important to choose hardware that is compatible with the collaborative program. Figure 4.2 illustrates a common networking diagram used in remote auditing. All devices are logged into the hosted collaborative meeting room. The iPhone should only join at the time of the production floor auditing; Wi-Fi boosters on the production floor help to keep the video and audio streaming during the walk-through.

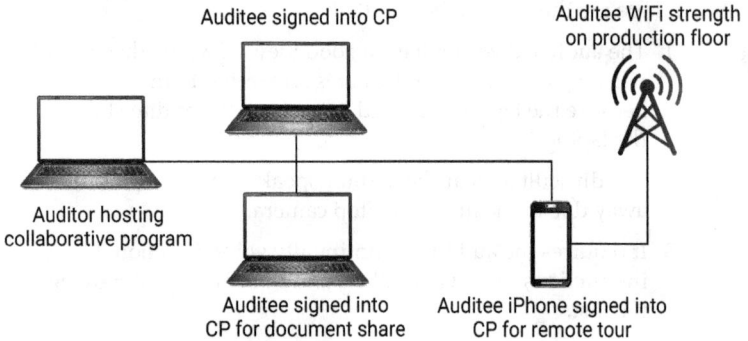

Auditee signed into CP

Auditee WiFi strength on production floor

Auditor hosting collaborative program

Auditee signed into CP for document share

Auditee iPhone signed into CP for remote tour

Figure 4.2 Remote audit network diagram.

INTERNET SPEED

Selecting the right internet speed that supports the necessary bandwidth requirements for streaming video, VoIP, and sharing documents in virtual meeting rooms is important. Faster internet will ensure these requirements. Before each remote audit, the Wi-Fi should be checked for bandwidth and boosted as necessary to ensure Wi-Fi is at its full capacity.

Collaborative Rooms

A collaborative room is used to display the streaming video over the internet from a computer or smartphone. There are more options for collaborative programs available to users. It is important when choosing to consider the tools necessary for remote auditing, how well the program will interface with your hardware, and the security features provided. At a minimum, the collaborative program should be encrypted or allow the user to password-protect the meeting.

VIDEO STREAMING

Video streaming will increase the internet bandwidth load and cause delays that will affect video and audio quality. *One option is to close other programs using internet bandwidth (like email) and encourage the virtual room attendees to do the same.* Be sure not to close the collaborative meeting program, however. Another option is to have the attendees using VoIP call in during the visual review to ensure that the video streaming keeps pace. If delayed streaming persists, then consider reducing the number of people in the virtual room (many people online in the virtual meeting room will drain bandwidth) by ensuring that only those needed are attending. In some countries, the internet bandwidth is poor, and it will be difficult to conduct a visual review or impossible to invite multiple attendees.

> *If video streaming is causing delays and affecting video quality, close programs using internet bandwidth, such as the camera, or use the phone line to call in.*

The collaborative room layout in Figure 4.3 illustrates the most used functions. When conducting remote audits, you will use all these functions, including videoconferencing and sharing the working area (or sharing your desktop) and the toolbar, to manage the virtual room.

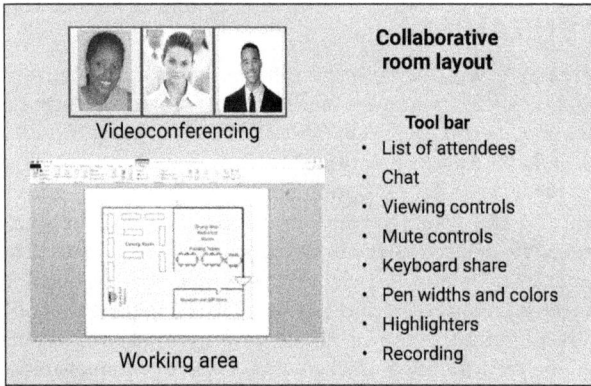

Figure 4.3 Collaborative room layout example.

MAPPING THE AUDIT TRAIL

Capturing the Environment

A satisfactory visual review of the operations can be accomplished in three stages. First, use the *factory tour* for an overview of the whole location. Next, use a regional view and focus on the *upstream and downstream work areas* to understand the workflow affecting the specific operation under audit. Finally, in the third view, *focus on the specific operation,* including processes, tools, job aids, and, if applicable, associated calibration stickers.

Following are the five steps in preparing for the visual reviews:

1. Plant tour: Have the auditee send you a map of the location. On this map you will need to let the auditee know what areas you want to review. Ask the auditee to define the associated location's restrictions and requirements. Such restrictions include sensitive areas that have restricted access, nonconforming product and part locations, electrostatic sensitive discharge–restricted areas, spare part storage, and cleanrooms, to name a few.

2. It is difficult to view any area when someone is walking and moving the camera at the same time. Items pass very quickly and may be blurred or delayed if the internet is not keeping up with the movement. To work around this, define the camera placement location on the map with an X. This way, the auditee can walk slowly to each camera placement location and then stand stationary while gently moving the camera for a panoramic view of the area. When walking to the next camera placement location, the auditee should keep the camera facing forward, not turning side to side. (If you see something on the way, you can always ask the auditee to stop and show a closer view of an area.) Number the placements so that you can follow along with the map, and everyone understands the location reference.

3. Plan for a *sound test meeting* about a month before the remote audit. The objective of this meeting is to check the technology and sound quality, ensure that use of a camera is approved by management, define camera placements, and review the audit plan. This is also a suitable time to communicate audit meeting rules so the audit sponsor can relay any necessary information to the audit team. Figure 4.4 shows a checklist to be used in planning and executing the sound test meeting.

4. Decide who needs to attend the *technical check meeting*. The person who is sponsoring the collaborative meeting room is the driver. If the auditee's collaborative meeting room is used, then at minimum the auditee sponsor and auditor need to join because the test ensures that the auditor's equipment works with the collaborative room and that it contains all the functions needed for the remote audit. (Of course, anyone within the auditee's company who has not used the company's collaborative room could join as well.) If the auditor's meeting room is used, then all personnel who participate in the audit should also be included in the *technical check meeting* (more detail later in this chapter).

Item	Actions prior to **Technical Check Prep Meeting**	Follow-up comments
	Log in ahead to complete any downloads:	
	• Ensure all auditees have e-mail invites and meeting room URLs	
	• Download program software prior to the remote audit. Check the company policy for camera use is approved	
	Technical Check on the Floor	
	Test Mobile Device:	
	• Check WiFi strength; Define low Wi-Fi areas on floor diagram	
	• Define area lighting (define low-light areas)	
	• Exchange cell phone numbers for immediate messaging access during the audit	
	Technical Check on Devices	
	Smartphone/tablet tests only:	
	• Smartphone ear plugs and microphone prepared (ensure clear communication between auditee and auditor during visual review)	
	• Exchange cell phone numbers for immediate messaging access during the audit	
	• Message a photo to the auditor's smartphone (use when sticker or document cannot be read while streaming)	

Continued

Figure 4.4 Technical check preparation checklist.

Item	Technical Check Area Designation	Follow-up comments

	Map the areas for audit:
	• Plant layout diagram
	  – Panoramic views (1, 2, 3) with direction indicators
	  – Camera placement location
	  – Regional areas
	  – Station views
	  – Nonconforming product location(s)
	• Highlight electrostatic discharge (ESD) areas
	• Material staging
	• Wi-Fi blackouts	

	Provide meeting rules for how to manage the microphone and cameras: • Mute the microphone during breaks • Stop sharing the camera when not in use • Ask if a closing meeting recording is needed	

Figure 4.4 *Continued.*

5. Use the whiteboard function within the collaborative program to work with the auditee to define the camera placement, audit path, and restricted areas (see Figure 4.5). This is a great time to meet the auditee and build trust.

Following are instructions on using a whiteboard:

- Depending on the whiteboard, either copy and paste a floor map onto the whiteboard or open the floor map file and select "Screen and keyboard share" for all to see and draw on.

- The auditor and the auditee will need to select a pen color to differentiate between markings.

- In Figure 4.5, a highlighter was used to capture restricted areas. An X indicates where the auditee should stand for the panoramic views. Now, get agreement on the camera placement location with the auditee to ensure clear visibility.

- With a circular line, draw the panoramic view, and number the views if there are multiple ones to ensure clear communication. Use arrows to let the auditee know in which direction you want the panoramic view to start and finish. Always move the camera away from a brightly lit area, like a window, to decrease glare.

- Define the station and associated regional view. Write "station view" on the map for clarity. If numerous stations are under review, consider a numbering format to align the overview sequence (3.1, 3.2). Ensure that the auditee has Wi-Fi at these locations; highlight areas with no or low Wi-Fi that could disrupt the audit.

- The auditor and auditee should print the final version of the map for reference during the tour and station review or save a copy of the highlighted map and email it to everyone in the audit.

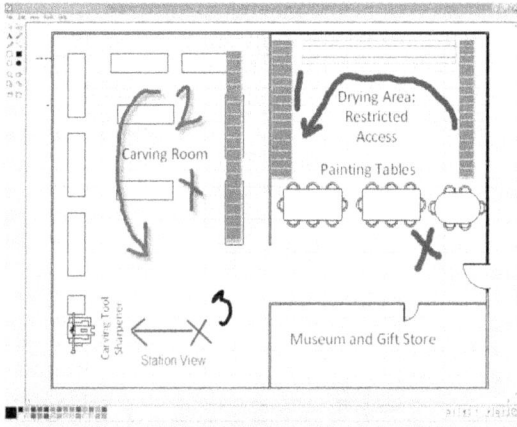

Figure 4.5 Numbered path used to tour facility. X marks the place to stand for panoramic views.

FRAMING THE VIEW FOR THE OPERATIONAL FLOW (REGIONAL VIEW)

Capture the Process Flow (Input > Operation > Output)

According to the map, go to the first *regional viewing*. For optimum viewing, instruct the camera operator to *keep the camera level and move slowly*. If you need to see something more closely, stop the auditee or camera operator, and ask him or her to walk slowly toward the subject of interest. If you need to read something and the focus is not sharp enough, have the auditee *take a photo of the text* and either email or message it to you. There are *two methods to use for an operational view*. The *panoramic technique, as used in the tour, or a side-step approach*.

Panoramic View

Define the camera placement location on the map for the operational flow. Slowly starting from the input station, move the camera to the station of interest, then to the output station (see Figure 4.6). Allow time to view the process flow before moving to the next station. Film around objects—it is better not to move items from the workstations while viewing. It is more important for the auditor to see how things really are. If the image seems dark in the camera's viewfinder, ask the operator to slow the camera movement and allow the light to catch up with the camera.

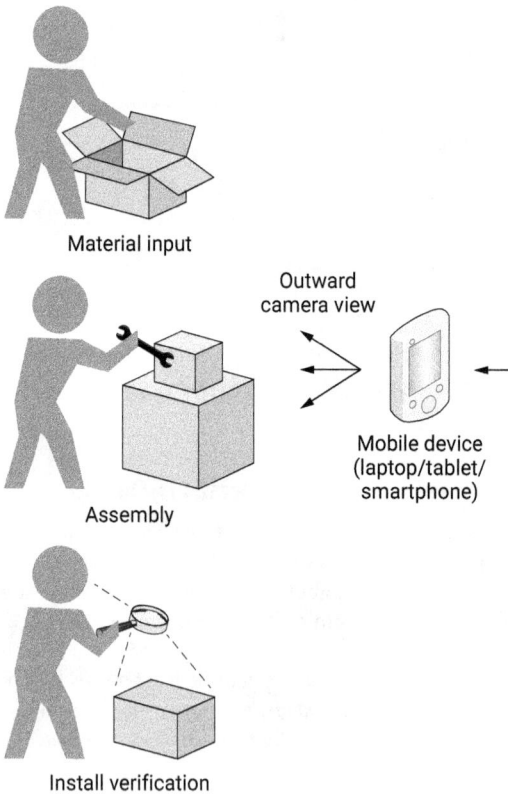

Figure 4.6 Panoramic operational review technique.

Side-Step Approach

Another method that could be used is a side-step approach. This method is not walking down the line with the material flow, but rather slowly side-step, wait, and step again (see Figure 4.7).

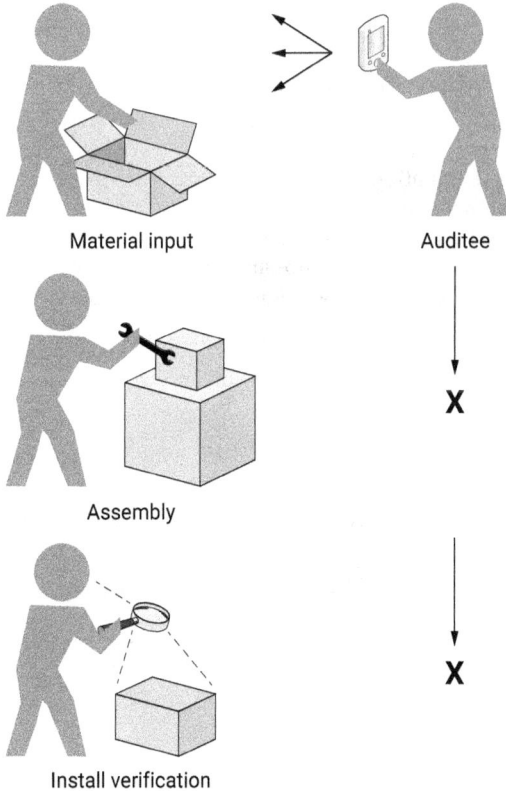

Material input

Auditee

Assembly

X

Install verification

X

Figure 4.7 Side-step operational review technique.

Slow camera movement is best to capture the overall process. Walking will move the camera too fast to catch what is occurring.

Define what method you prefer the auditee to use before the audit, so they know what is expected. It would be great if they tested this ahead of time and could see for themselves the issues associated with fast camera movement.

Station Detail

To capture a whole station, stand back from the station for a wide-angle view, then walk inward to the station for a closer view, avoiding a fast-sweeping view of the station. To view small text on travelers or stickers, hold the camera stationary to capture the text. If the text is unclear and you are unable to read the text, have the auditee take a picture of the record and email or message it to you. Photos can be clearer than streaming video. If this is not possible, get the documents scanned and emailed to you. Look for these items where applicable (see Figure 4.8):

1. Documentation like visible job aids and procedures

2. Tools, including their condition and possible poka-yoke solutions

3. Station layout and configuration are organized, and the station is clean; verify calibration stickers

4. Material review for proper handling, storage, and product traceability

Job aid: Provide the auditee with diagrams such as in Appendix A showing how to handle the camera and visual angles.

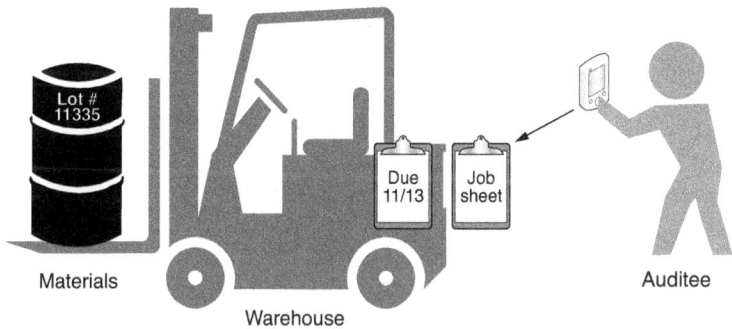

Figure 4.8 Station detail.

CHAPTER REVIEW

- The three process elements that need to connect for a successful remote viewing are a mobile device, Wi-Fi, and a collaborative program.

- Equipment checks include equipment, software, location mapping, and camera practice.

- A map is a guide the auditor uses to understand requirements like ESD, and process flow, and to direct the auditee on where to place the camera for viewing.

- Collaborative whiteboards are easily used and highly interactive, allowing all parties to draw a walking path, highlight good stopping points for wide-angle views, and mark restricted areas.

- Moving the camera or presenting a side view while walking makes it difficult to see things because the motion is too fast. Walk toward objects being viewed and stand in one spot for a panoramic viewing.

Chapter 5

Audit Models to Consider

L earn to define the best audit model that replicates the working environment, meets an immediate need, and supports the audit process:

1. Why you would choose to conduct virtual audits

2. What different audit models can be applied to best meet quality objectives and business needs

3. Review the remote audit process

Knowing how to choose the best audit method is critical to the success of the audit. Match the auditee's working environment as closely as you can to best understand their experiences, processes, and organizational communication.

Things to consider when determining the best audit method are:

1. Will this audit replicate the working environment? Does the workforce do business over the computer or only face-to-face?

2. How is organizational communication delivered? Are broadcasting messages distributed throughout a remote workforce or do workers attend on-site "all hands" meetings? (See Figure 5.1.)

3. Are records located online or stored in hard copy?

4. Is the business in a safe location in the world, free of governmental travel warning alerts?

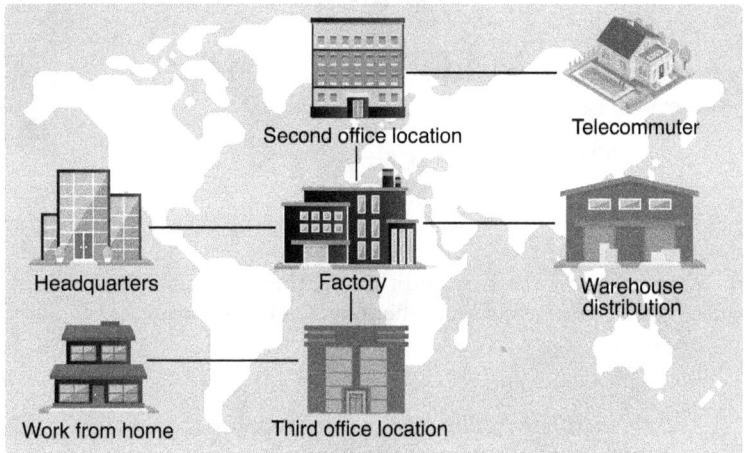

Figure 5.1 Remote workforce distribution example.

5. If remote, would a face-to-face audit be timely? Is there an immediate need to review a factory or supplier's process?

6. Do company budgets prevent travel, adding challenges to an audit schedule?

Technology should be used as an auditing tool, *making audits less likely to be cancelled or postponed due to travel issues or costs. Technology should be used for more oversight where it is needed.* Rather than thinking of auditing as only a face-to-face, on-site process to gather data and test procedures, auditing can be accomplished virtually (remote audits) or in a combination of part face-to-face (on-site) and part remote audits (remote) called a *hybrid*.

When the auditor prepares the audit checklist, they should consider the audit model (face-to-face, remote audits, or hybrid) and the associated methods. The checklist and data collection methods will be affected by what is being audited remotely versus on-site. This section will discuss the different audit models, processes, and policies.

TRADITIONAL ON-SITE AUDITS

We are all most familiar with on-site, face-to-face, traditional audits. Traditional on-site audits are best to consider when a business is colocated—where all employees and work environments reside in a single location—especially when the internal communication processes exclusively rely on a paper system and face-to-face meetings. If you do not audit your own process, *colocated* businesses have been highly successful using face-to-face auditing.

Another reason for a face-to-face audit would be escalating supplier issues resulting in the need to visit the supplier's operations. Good reasons for traveling to a supplier operation would be to see firsthand the operations and meet in person with someone you may have only spoken with on the phone. These are all very good reasons to use an on-site, face-to-face audit. Table 5.1 reviews the strengths and weaknesses of face-to-face audits. Can you list other strengths and weaknesses of your situation and your environment?

Table 5.1 Strengths and weaknesses of the face-to-face audit model.

Face-to-face model characteristics	Strengths	Weaknesses
• Domestic or international • On-site face to face • Manufacturing floor or operations • Hard copy and/or paperless • Monthly all-hands meetings to disperse business information • Auditor is escorted through factory and offices	• Face-to-face provides the ability to view nonverbal communication • More free to review a factory floor or operations • Face-to-face audits are the best to use in cases of escalating issues or very high-risk	• Computerized systems can be overlooked during the audit if only hard copy records and manuals are provided during the face-to-face audit review • Delays to a supplier site visit due to cost and schedule • Additional travel expenses and labor costs can deter proper oversight

ALTERNATIVES

What if the colocated business chooses to conduct timely supplier remote audits and catches emerging issues before they escalate into ones that require an on-site supplier audit? Wouldn't that be *more proactive in prevention and more cost-effective?* Remote auditing will not replace face-to-face audits, but it does enhance the objective of auditing. When audits are postponed or cancelled during economically challenging times because of travel expenses, remote auditing should be used to *stay on top of supplier oversight*, even when a business is colocated.

REMOTE AUDITING

Remote audits can even be performed internationally from home offices. Remote auditing depends on technology to view operations using cameras and the internet to connect to a collaborative meeting room. Special camera handling procedures help to create clear views for the auditor to verify processes and operations. Screenshare functions allow the auditor to see records and documents and associated traceability much more clearly than in person.

Because auditors cannot depend on non-verbal communication, auditors must learn to ask more clarifying questions to validate the processes. Often, too much reliance on nonverbal communication for answers to questions can lead auditors in the wrong direction or cause them to discontinue further questioning. In some cases, remote auditing is better than being there in person. Table 5.2 reviews the strengths and weaknesses of remote audits.

HYBRID AUDIT—COMBINED REMOTE AUDIT AND ON-SITE AUDIT

Hybrid audits occur when there is a need to conduct a portion of the audit on-site while another part can be accomplished through a remote audit. Rotating remote auditing with on-site audits annually could be a viable option for many businesses. On-site locations may not have available Wi-Fi access on factory floors, which would prevent the use of cameras.

Table 5.2   Characteristics, strengths, and weaknesses of remote audits.

Characteristics	Strengths	Weaknesses
• Multiple locations either domestic and/or international	• Meets an immediate need to review a remote process	• Will not work if Wi-Fi or broadband internet is not available
• Service, business, and/ or manufacturing processes	• See online records clearly and watch how the record is located	• Slow network speed prohibits the use of cameras
• Online record reviews	• Can level the playing field without visuals; silent types or physically challenged people say their voices are more likely to be heard	• Limited visibility for nonverbal communication
• Visual reviews (with the use of cameras) conducted over the internet	• Surprise audits are more adaptive and agile	• Lack of skill and experience in technology

Initially, adopting a hybrid audit model may facilitate acceptance of the method by people just learning how remote audits are conducted (see Figure 5.2). As a business proceeds using these advanced tools for auditing, participants will become more comfortable and experienced with remote auditing. When beginning to use remote auditing techniques to sample and verify components at the next on-site audit. Try to understand the different information your remote audits and on-site audits provide.

When implementing a remote auditing program (see Appendix C checklist) you may experience the following:

- *People typically will disclose more information* while talking to you online than during on-site audits.

- *Camera viewing may be limited compared to seeing the operation in person until you set up* specific viewing instructions, new viewing stances, and lighting changes to improve results for a remote audit.

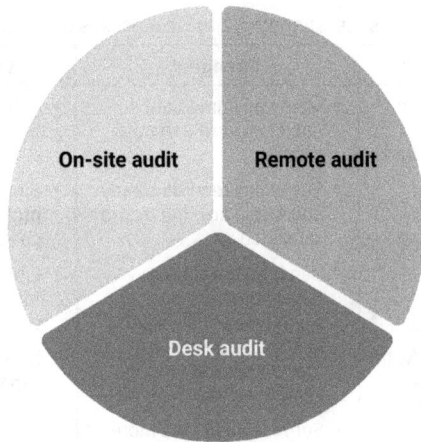

Figure 5.2 Hybrid audit model.

- Expect technical difficulties or *lack of technical skill level to occupy more energy than the* remote audit *and* possibly create frustration.

- The remote audit *report will contain more information than the on-site audit report* because of the continued access to the computer and the improved visibility of records, documents, and, in some cases, the equipment on the factory floor.

> *People typically will not disclose as much information in face-to-face audits as in remote audits.*

Once you become experienced in questioning techniques and how to remotely operate a camera, conducting remote audits in distant locations such as suppliers will become more routine.

When integrating remote and on-site methods, first conduct the remote audit. If there are instances where you were not able to collect objective evidence due to physical limitations, sampling issues, or high-risk activities, a subsequent on-site auditor can collect the necessary evidence. In other cases, you can arrange for a qualified auditor who is in the region to visit the auditee site to witness the evidence. This person could be called a *proxy* or *surrogate auditor*.

Table 5.3 reviews the characteristics, strengths, and weaknesses of hybrid audits.

Table 5.3 Characteristics, strengths, and weaknesses of the hybrid audit model.

Characteristics	Strengths	Weaknesses
• Multiple locations • Domestic/international • Includes on-site manufacturing audits with manufacturing process inputs • Includes virtual business process audits with business/ service process inputs • Can alternate audit types each year • Can integrate or combine remote and on-site audit methods using a proxy auditor	• Initially, using a hybrid audit model will help to define remote auditing processes • People respond more favorably to a hybrid audit model than to an exclusive remote audit model • Use of both auditing models enables a better understanding of the whole system (physical and online)	• Over time, the travel expense and labor costs may result in canceled on-site audits • Auditee personnel may favor one approach over the other and express their biases or become uncooperative • Alternating face-to-face and remote can cause confusion

REMOTE AUDITING PROCESS

When different time zones (such as a 12-hour difference) become an issue in remote auditing, conduct the opening meeting with a person who has the authority to approve the audit plan.

Send meeting invitations to all audit participants to schedule them for their audit using the collaborative room's invite tool. This way, the auditees can either accept or decline the time frame and download the collaborative meeting software before the audit. Typically, schedule a one- to two-week period to conduct a full system audit. Audit interviews take place over two hours at the best time in the auditee's schedule. When the interviews and audit report are completed, contact the same authorized person and review the findings in a closing meeting. Record that closing meeting with the approval of the auditee, using the recording tools in the virtual meeting room (see Figure 5.3). Also, as with the opening meeting, you can give the participants keyboard and mouse controls to enter their names, and then the auditor can use the same form to record meeting minutes.

Remote audit process considerations:

1. For the opening meeting, schedule a time with the audit sponsor or area manager to review the audit scope and objectives and to confirm an initial closing meeting date. *Agree on the scheduling method and collaborative meeting program.*

2. When arranging meeting times, the auditor provides the manager with open times for the auditees to make arrangements.

   • *Ensure that the collaborative program provides all the functionality needed for the audit (including recording the closing meeting). (Note: The collaborative program used will define who sends the meeting invites and reminders because of collaborative program access.)*

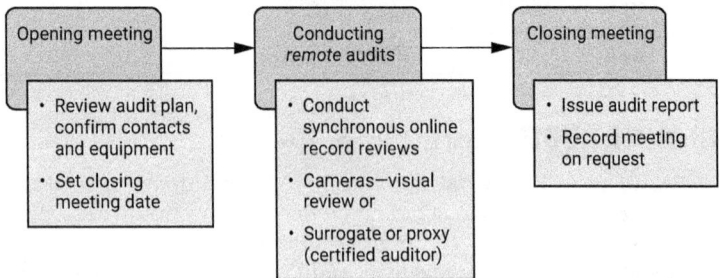

Figure 5.3 Remote auditing process model.

3. *Define the technology needed.* For example, if a camera will be used as part of the audit, *cover the logistical details* to ensure the correct resources are in place for the audit, such as IT personnel or assistants.

   Conduct the appropriate audio, video, and functionality testing.

   - *If a surrogate, proxy, assistant, or other auditor is used instead of a camera, or there are other specific audit evidence issues,* ensure that he or she completes all required documents to enter the place of business and has management approval as applicable.

4. *Conduct the remote audit using the online collaborative program and desktop share functions for record reviews, and cameras or surrogates for visual reviews, as applicable.* (See Figure 5.4 and refer to earlier chapters for details.) Update your audit report with findings and comments at the close of the audit to eliminate having to remember interview statements over an extended period collectively.

   - It's best to do all the camera location work at the same time because gearing up for camera work can be very time-consuming. For efficiency, when the floor tour is complete, go to the scheduled area to conduct the visual review rather than returning later.

   - Use a camera to verify items not captured with online records.

   - Photograph the record or scan the records of items not visible during a visual review (cameras) for the online record review.

   - If Wi-Fi does not reach the area, have photos taken at that time to share upon return to the record review.

5. When the remote audit is complete, conduct a final review of your audit report before meeting with the sponsor. *In the closing meeting, the auditor can show his or her screen to allow all to view the audit report as the findings are discussed.*

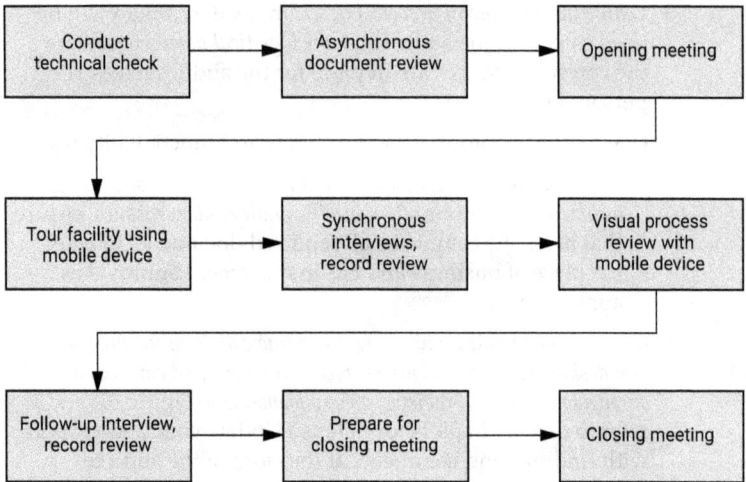

Figure 5.4 Remote auditing process steps.

This gives the sponsors and auditees a visual and audio learning experience for improved communication. If the sponsor requests that the closing meeting be recorded for absentees, ensure that the collaborative meeting room has a recording functionality.

> *If you want to conduct a remote audit, but a camera cannot be used or has limited access, use a surrogate, proxy, or assistant to witness objective evidence.*

BUSINESS POLICIES

Auditing organizations or departments using virtual auditing models as tools in their auditing program should also define business policies that help to provide direction (see Appendix C for suggested documents). Here are some areas to consider:

- *Escalation guidelines* for remote auditing might be needed. For example, a continually postponed online audit could become an emerging issue that might result in an escalation that is better suited for a face-to-face approach. Define the parameters that best fit your business culture and objectives.

- When surrogate or proxy auditors are used, the organization should *define minimum auditor credentials and identify or qualify the organizations* that can provide auditor resources. For instance, the American Society for Quality Excellence (ASQE) offers auditor certifications. Visit www.asq.org/cert. Also, the IRCA (IRCA.org) or RABQSA (RABQSA.com) has a directory of qualified auditors that can be found online under "find an auditor by state and country."

- *Define policies for each audit model type* that could be used in your business. Document the audit processes and standard equipment to use, and train and certify your internal auditors (for standardization).

- *Connect your global supply chain* using the different audit models. Even if your organizations are under different management structures, have your auditing teams link the company processes to ensure continuity and avoid redundancy.

CHAPTER REVIEW

- Areas to consider when choosing the best audit model for your business include:

  A. Work environment

  B. Expenses

  C. Hazards and safety concerns

  D. Organizational communication practices

- The three models discussed in the chapter are traditional on-site, remote audits, and hybrid models.

- If some participants cannot attend a meeting, it could be recorded to provide accurate and consistent feedback for those not available at the time.

- Techniques to verify remote processes include:

  A. Audit assistants, surrogates, or proxies

  B. Cameras

  C. Photographs taken at the time of the audit

Chapter 6

Validation Challenges and Remote Audit Risk

U nderstand important process criteria to validate your remote audit:

1. Discuss validation definitions

2. Validation issues addressed

3. Process and methods that support remote auditing validation

4. Remote audit risks

The definition of a remote audit should be the same as that of
an audit in which on-site or direct means are used to collect evidence.
If this were not true, a remote audit would not be an audit.

—J.P. Russell

DEFINITIONS AND VALIDATION CHALLENGES

Several challenges are presented by this emerging method for conducting audits. A source of concern about remote auditing comes from a *lack of experience in conducting audits online or in working on a virtual team.* Many of the issues are caused by a lack of *understanding and training on how to apply the audit process* using the available technology. Because it is a remote audit, there may be some *feasibility issues* in certain environments.

Can a remote audit be validated by conducting the same audit face-to-face and then comparing results to see whether the remote audit found the same nonconformities that the face-to-face audit did? If the remote audit found more nonconformities, would that make it a better audit than the face-to-face one? What does it mean if the two audit methods find different things?

Let's begin with definitions:

> *According to the International Organization for Standardization, an audit is a systematic, independent, and documented process for obtaining audit evidence and evaluating it objectively to determine the extent to which audit criteria are fulfilled (ISO 19011:2011,* Guidelines for auditing management systems*).*

A definition for remote auditing by J.P. Russell (active in international standards development, standards, and auditing) is below. This definition could be synonymous with remote or online audits:

> Remote audit: *An audit (see definition above) that uses electronic means to remotely obtain audit evidence to determine the extent of conformity to the audit criteria.*

We should expect the definition of an audit to apply to audits conducted remotely as well as on-site audits. One of the critical concerns, when data collection is not sufficient, is that the audit credibility can be compromised (J.P. Russell remote auditing paper, 2009). This means we should expect no less from remote auditing than we do from traditional auditing. As audit professionals, we should require that remote audits be systematic, independent, and use the same basic process as face-to-face audits. These expectations cover opening meetings, delivering on audit objectives, collecting data from interviewing, inspection, verifying documents and records, counting, measuring, and observing tasks, plus process inputs and outputs, and ending in formal closing meetings. The rub when it comes to remote auditing is how we collect data and verify them using the internet and other technology. The evidence collected from an on-site audit may be different than the evidence collected from a remote audit due to using different paths or strategies for conducting the audit.

Q&A WITH THE LATE J.P. RUSSELL

In the January 2011 issue of *Quality Progress* magazine, the late J.P. Russell wrote in his column, "Standards Outlook," about remote auditing and titled the article, "Remote Control."

Q: Are audits conducted remotely still audits, or are they surveys?

A: That depends on the process and standards used. If a remote audit follows the audit process, seeks objective evidence, and can validate the source, I would call that an audit.

Q: If the auditor never sets foot in the area being audited or interviews personnel face to face, can audit objectives be accomplished?

A: Yes, my experience has been that remote workers do not meet face-to-face with their peers/colleagues, yet work objectives are accomplished. I believe that in these cases auditing remotely enables a better understanding of the new, emerging virtual work environments. Remote auditing also allows the auditor to get answers to the hard questions about communications. The answers to these questions are often assumed in a face-to-face interview.

Q: If a potential supplier claims to conduct remote audits, should you be concerned?

A: I would first ask to see how its remote audits are conducted and how the supplier validated the evidence. My concern levels will either rise or fall based on their answers.

Q: When auditing a corporate global supply chain, how can we still believe site audits will capture the systematic processes if all the processes are not part of that site audit?

A: I don't believe site audits can capture a systematic view of a corporation's supply chain. I think it is naive to even believe a combination of site audits will capture the systematic view in a global certificate. The attributes of a global supply chain change the scope of an audit (not by location, but by processes). To best match process and workforce environments, a remote audit or hybrid model should be used.

> *A remote audit is an audit that uses electronic means to remotely obtain audit evidence.*

NO PHYSICAL PRESENCE BY THE AUDITOR: TOUR THE AREA

An auditor may tour an area to collect information. The information may verify conformity/nonconformity or aid the investigation. In many cases, touring the area or making on-site observations are necessary to ensure there is sufficient audit evidence. This is especially true for processes that involve skills (for example, sawing, labeling, cleaning, or sorting) and/or physical products, equipment, or materials. Physically touring areas that do not involve people skills, product, equipment, or process materials yields little new information, so these areas—normally processes that are administrative or service-related—are well suited for remote auditing.

Some examples of areas that may lend themselves to remote auditing tools are:

- Document control

- Planning

- Management review

- Training

- Corrective/preventive action administration

In remote audits, it may be challenging to make objective observations remotely. One must consider the availability of impartial camera (video or still pictures) coverage. For some situations, it may be necessary to employ a proxy or surrogate auditor to make on-site observations.

The need to tour or observe an area also depends on the purpose of the audit. If there is a safety audit of the document control or administrative offices, people may need to demonstrate certain skills (using a fire extinguisher), show that they have access to equipment, and show that the equipment has been maintained. For low-risk operations (controls), use of video equipment to make real-time observations may be justified. The video equipment could be provided by the auditing organization or auditee organization. In other cases, existing surveillance cameras may be all that are required to verify conformity/nonconformity.

The remote audit program management must consider the challenges to ensure that remote observations are sufficient and representative.

What Has Changed Now to Make Remote Auditing a More Appropriate Auditing Tool?

First, technological advances in mobile devices allow for clear and concise viewing at a remote location. Second, we are adapting to using technology. The cultural shift to self-service in customer services using the Internet or chatbots has taught the public how to use online tools, which has opened the potential for remote audits. The collaborative program has taught us about videoconferencing as we learn to make sharper and clearer images of grandchildren or our gardens or to share the latest home improvements with distant friends and loved ones.

As a society, we are going paperless. Many, if not most, use online banking and invoicing, and read magazines on a tablet. And product/service development and marketing are moving from a push strategy to a pull strategy that requires more interaction by consumers. Our lives are becoming borderless, and so should the way we audit our processes.

AUDIT RISKS

Audit Risk: Conducting the Audit

Many of the same risks exist for both remote audits and on-site, face-to-face audits. For remote audits, risks associated with travel are negligible. For example, for remote audits, the risk of not finding the audit site, getting lost, or not arriving on time at the physical location is not an issue. For face-to-face audits, when auditors travel long distances, a travel day may be added to ensure that transportation delays do not impact the audit schedule. Adding extra travel days is expensive and inefficient.

There are aspects of remote audits that could pose a significant risk. The level of risk may depend on the scope of the audit, the relationship between the auditing organization and auditee, the criticality of the product or service, and the cultures and nature of the organization.

The auditing organization and auditor should identify aspects of the remote audit that represent a risk to the audit. Once identified, they should be assessed, and then significant risks should be analyzed and treated or avoided.

Some potential significant risks for remote audits are:

- Communication equipment fails or is unreliable.

- Internet access fails or is unreliable.

- Internet security—information confidentiality and integrity.

- Auditee personnel do not adhere to scheduled meeting times.

- Auditee personnel constantly leave scheduled meetings to take care of other business.

- Auditor accepts audit evidence that is not objective and impartial.

- Auditee stages certain events or activities of the audit.

- Auditor or audit program manager approves an on-site proxy auditor who is not qualified to expedite the audit.

Remote Audit—Pledge of Cooperation

I understand that when audits are conducted from a remote location that cooperation and civility are critical to the successful completion of the audit. I promise to be on time for all scheduled meetings and events, to cooperate to the best of my ability, and that I will inform audit leaders of any changes that may impact the audit.

Audit Risk: Cooperation and Communication

Many of these same risks exist for on-site audits, but due to the virtual nature of remote audits, there may be higher levels of risk that need to be addressed. For example, in some cases, an auditor proxy or surrogate can be used to alleviate some high-level risks associated with collecting audit evidence.

One possible way to lower the risk of schedule delays is to ask auditee personnel to sign a pledge of cooperation before the opening meeting. This should not be a big deal but will highlight the importance of adhering to schedules. In the virtual world, an auditor or audit coordinator cannot walk down the hallway or find the person in the break room to remind them the meeting was supposed to start five minutes ago.

If signing a pledge is not practical in your situation, the lead auditor can request a verbal affirmation at the opening meeting.

Audit Risk: Security

Internet security risk can be minimized by setting policies with auditors as to how they access the internet. For instance, free public Wi-Fi should never be used when auditing. As an added protection, hardwire connection to the internet from your computer at a minimum ensures that the internet connection is password protected. Ensure that virus protection is current. Use the companies' collaborative programs or a reputable collaborative company like GoTo.com, MS Teams, Zoom, or Google Workspace when meeting online. Auditees do not need to fear that auditors will take over their files because auditees always stay in control of their files.

TECHNIQUES TO MITIGATE AUDIT RISKS

For remote audits, it may be necessary to go on-site to make observations and/or transmit documents and records. For remote internal audits, it could be someone in the auditee organization assigned to assist the auditor. The person selected should have the minimum conflict of interests and be known to be trustworthy (perhaps someone from the accounting department).

An assistant could be assigned for external audits, too, but they may need to sign a statement that they will follow the auditor's instructions in an ethical and unbiased manner. Most people are not going to tamper with information because it is a dismissible offense. In other cases, you may be able to hire what I have called a *surrogate* or *proxy*. The surrogate or proxy auditors could substitute for the external auditor for specified tasks such as making observations, selecting records to inspect, and so on. Surrogates or proxy auditors must have the appropriate competencies. Both the auditing organization and auditee may seek surrogates or proxies to benefit from remote audits.

REMOTE AUDIT CONSIDERATIONS

Please consider the following when planning, implementing, or reviewing remote auditing techniques:

- Travel: could there be significant savings in travel costs and employee productivity?

- Audit scope and objectives/purpose: new certification, routine, supplier approval, system, or process audit.

- Nature of the processes to be audited: does the process to be audited involve only oral communications and/or recording (documenting), such as order entry, purchasing, and document control.

- Type of product, equipment, and materials involved in the process to be audited: Is the equipment used to conduct the process (drilling, cleaning, or inspecting)? Is observing the operation of the equipment, moving of products, or handling of materials a critical factor for verification of conformity/nonconformity (compliance/noncompliance)?

- Number of locations: the more locations (terminals, branches, offices), the more an organization would benefit from remote auditing techniques.

- Internal or external (corporate, supplier): Internal audit risks would be lower than for external audits. External audits require a higher standard for collecting sufficient and representative data.

- Availability of appropriate electronic communication equipment.

- Capability of the auditor and auditee to operate electronic communication equipment.

- The need for security must be addressed.

- Implement an auditor training program to ensure the success of the program. Consider auditee needs, too.

CHAPTER REVIEW

- The following are some issues that may be encountered during the implementation (see Appendix C for implementation checklist) of a remote audit program:

    A. An auditor's lack of experience conducting audits online and collaborating with virtual teams.

    B. An auditor's and auditee's lack of training on how to apply virtual techniques.

    C. Not recognizing feasibility issues in certain environments.

- A remote audit is when electronic means are used to remotely obtain audit evidence to determine the extent of conformity to the audit criteria.

- The evidence collected from an on-site audit may be different than that collected from a remote audit because different paths or strategies were used for conducting the audit.

- The need to tour or observe an area may depend on the purpose of the audit.

- The remote audit program management must consider the challenges in ensuring that remote observations are sufficient and representative.

- The level of risk of a remote audit may depend on the scope of the audit, the relationship between the auditing organization and auditee, the criticality of the product or service, and the cultures and nature of the organization.

- Free public Wi-Fi should never be used when auditing.

- Conduct open discussions about issues and concerns as part of your training program.

Conclusion

This book intends to answer the question of how we *apply the audit process to audit remotely using technology and online communication.* We covered:

1. The benefits, important terms, challenges, and environmental driving forces of remote auditing

2. What communication theory is, and how to better communicate online

3. How to use the available technology to build trust and perform record reviews, and use cameras to inspect objects and verify processes

4. Audit models to consider as part of the audit process

5. Audit validation and risks

The essence of any audit is the credibility of the information collected—whether the data are objective, and representative of the sample population. This is true for both face-to-face and remote audits. There are good and bad on-site audits as well as good and bad remote audits. *If you do remote audits, you want to do them correctly to ensure they are credible.* Remote audits can be a welcome addition to your organization's options that will improve oversight and lower expenses.

Appendix A

Technology—Visual Tools (Illustration)

Job Aid:

Provide the auditee with the diagrams on the following pages, showing how to manage the camera and visual angles.

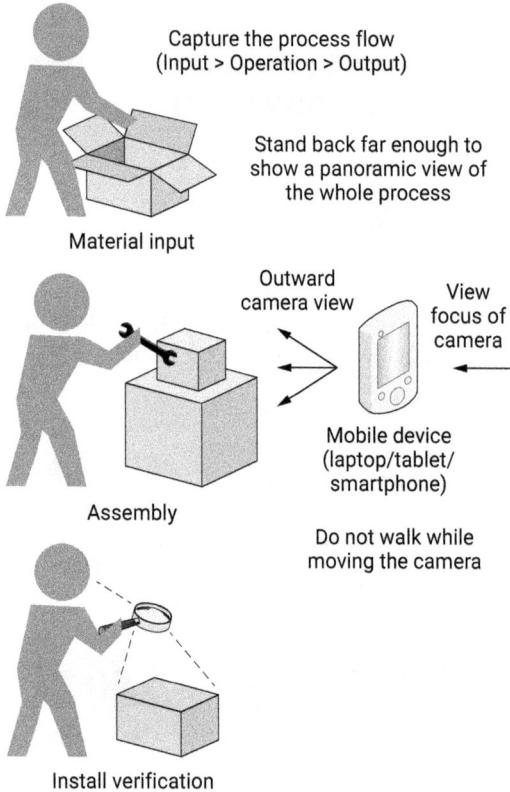

Figure A.1 Panoramic operational review technique.

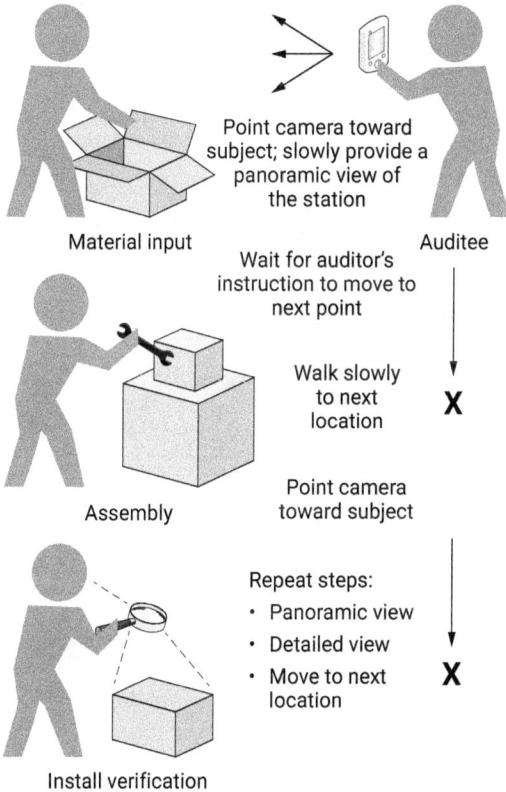

Figure A.2 Process flow side-step approach.

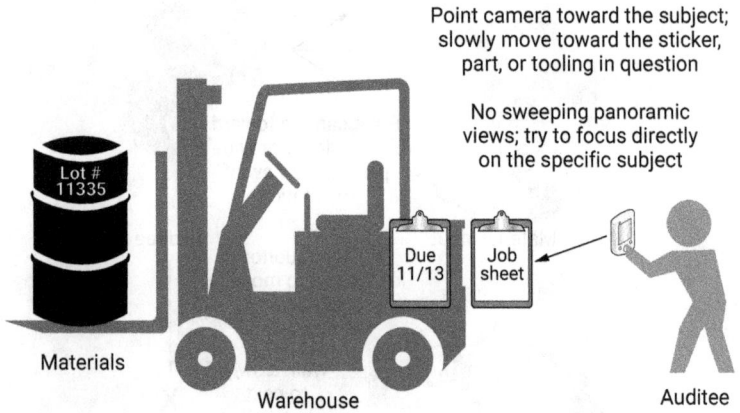

Point camera toward the subject; slowly move toward the sticker, part, or tooling in question

No sweeping panoramic views; try to focus directly on the specific subject

Materials

Warehouse

Auditee

Figure A.3 Station detail.

Appendix B

Risk Assessment
PFMEA Analysis

Process Function	Potential Failure Mode	Potential Effect(s) of Failure	Potential Cause Mechanisms of Failure	Recommended Action
Trust Building	Low monitoring behavior	Verbally hostile environment, lateness, delays due to texting, and interruptions	Lack of FtF monitoring for auditor and auditee	Auditor must have the skills to use technology while conducting an audit simultaneously (virtual meeting experience) and the ability to facilitate a virtual team and manage potential behaviors
	Uncertainty reduction theory	Audit may slowly or not proceed	Unskilled at keyboarding, use of technology, unable to find or access documents	Provide an audit plan early to ensure preparedness, conduct technical checks
	Missing non-verbal communication	Hostile environment	Using non-verbal communication for audit decisions	Build trust with remote auditees, be aware of behavior changes that can occur while auditing remotely, and use audit questions to validate evidence and replace reliance on nonverbal communication
	Use of camera/microphone	Privacy issue or unable to proceed	Missing permissions	Permissions for use of camera and associated handheld devices

Process Function	Potential Failure Mode	Potential Effect(s) of Failure	Potential Cause Mechanisms of Failure	Recommended Action
Tech-nology	Broadband performance	Failure to connect properly	Slow, unable to support VoIP, video	Network is secure and meets broadband requirements for all parties
	Collaborative software	Auditor is not able to use technology to find objective evidence	Lack of camera, VoIP, presenter assignment, list of attendees, desktop share, text, keyboard share, drawing capabilities	Ensure software contains all necessary remote auditing tools
	Technical difficulties	Poor visibility of objective evidence and poor sound quality	Bandwidth issues, configuration of computers and mobile devices	Technology checks should be used to determine Wi-Fi weakness, software and equipment performance prior to the audit
	Network security	Exposure to confidential records	No internet, no security software installed, or updates not up to date	Technology checks should be used to ensure internet security is secure
Method	Location mapping	Auditor does not understand the location when using mobile devices	Technical pre-check not conducted	Remote location mapping is provided; including specific areas noted: ESD, nonconforming materials, and equipment etc.
	Mobile device handling	Unable to clearly see processes, product, or stickers	Moving too quickly with the mobile device, walking and moving the camera simultaneously	Provide instruction for handling mobile devices to clearly see objective evidence

Process Function	Potential Failure Mode	Potential Effect(s) of Failure	Potential Cause Mechanisms of Failure	Recommended Action
Method (cont.)	Remote directing	Areas under review that are not shown	Remote person is confused about where they are supposed to go	Provide instruction on how to travel through the remote location, including stopping points for panoramic views for tours, specific processes and areas (non-conforming products) to review
	Collaborative SW tools	Poor audit, unable to collect sufficient evidence	Unfamiliar with facilitation tools	Training/experience facilitating virtual meetings
	Active vs. passive communication	When to talk, too much talking by a few, or silence from others	Expected meeting rules not set by the auditor	Set meeting instructions for timeliness, microphone and camera use, respect for others while talking, and copying materials
	Camera and/or microphone use	Lack of privacy		
Communication	Synchronous	Attendance issues	Audit planning and preparations did not address availabilities or different time zones	Limited attendance if audit is not conducted during regular working hours
		Lose access during the audit	Remote contact information not provided in preparation steps or lacking adequate cell phone reception	Provide alternative communication methods in case of technological difficulties

Process Function	Potential Failure Mode	Potential Effect(s) of Failure	Potential Cause Mechanisms of Failure	Recommended Action
Commu-nication (cont.)	Asynchronous	Access issues	Use email over central record storage, cannot find documents in master file	Provide audit plan early to ensure preparedness; but it could be a sign the document locations are not used, but email is

Appendix C

Remote Audit Program Management 10-Point Implementation Checklist*

No.	Description of Requirement	Done/ Status/NA
1	Is there agreement between parties on *which standards, documents, or procedures will be used* for remote audits? For example: ISO 9001, AS9100, ISO 14001, and so on. Is there *mutual agreement of cooperation* needed for virtual communications?	
2	Is there agreement between parties (client, auditee, and auditing organization) on acceptable *information security measures and safety issues* (e.g., electrostatic discharge)?	
3	Is there some type of *management plan* for the implementation and control of the remote audit program? Does it state *which remote audit model(s) will be used* for different situations and environments? Are audit program and individual *audit risks considered* (identified and evaluated)?	
4	Do procedures require that the *audit plan identify remote auditing techniques* (asynchronous and synchronous) that will be used during the audit, such as webcam, camera, collaborative software whiteboard, and recording? Does the audit plan include a technical check prior to the audit date?	
5	Do procedures require, or is there a record relative to, *auditors' abilities to understand and use information technologies* (i.e., document control, global work environment) *employed by the auditee* organization to manage its management system processes?	

* Taken from guidance in this book and ANAB Computer Assisted Auditing Techniques (CAAT) CB Application for Accreditation Guidelines, FA 2020.04, July 2012

No.	Description of Requirement	Done/ Status/NA
6	Do procedures require, or is there a record relative to, *auditors' abilities to understand and use information technologies that will be used as part of the remote audit process? Are auditors competent* (trained or experienced) in remote auditing techniques that include uses of technology, conducting meetings, interviewing, and addressing anxiety concerns?	
7	Do procedures or guidelines address how remote audit *time will be accounted for* as part of the individual audit process? For example: opening to exit meeting time is counted toward performing the audit time, time spent prior to the opening is planning and preparing time or, in some cases, for document review.	
8	Do procedures or guidelines require *that audit reports indicate the extent to which remote auditing has been* used in carrying out the audit and how it contributes to audit effectiveness and efficiency.	
9	Has the auditing organization provided the *necessary infrastructure and management support for a remote auditing program* and use of remote auditing techniques? Are issues involving *resistance to change* from face-to-face to remote audits addressed?	
10	Based on business/organization risks, is there a *plan to physically visit the auditee periodically,* such as every three years or five years?	

Glossary

assessment—A systematic process of collecting and analyzing data to determine the current, historical, or projected status of an organization (ASQ QP Glossary).

asynchronous system—*See* synchronous and asynchronous.

audit—The inspection and examination of a process or quality system to ensure compliance to requirements. An audit can apply to an entire organization or may be specific to a function, process, or production step (ASQ QP Glossary).

blended—Using both on-site (face-to-face) and remote (electronic) techniques.

colocated—To locate or be located at the same site (for two things, groups, military units, and so on).

Face-to-face (FtF)—The auditor and auditee are face-to-face. Auditing on-site would be face-to-face.

interactive—Exchange between the auditor and the auditee. Exchange of information orally, visually, or by text.

kinetic—Of or relating to the motion of material bodies and the forces and energy associated therewith.

Remote audit —An audit in which electronic means are used to collect audit evidence. Electronics may include computers, collaborative software (such as GoTo.com, Teams, or Zoom), mobile devices, microphones/speakers, or telephones. Sometimes also called *online audit* or *virtual audit*.

streaming video—Broadcasting live to the internet from a computer, tablet, or smartphone.

survey—The act of examining a process or questioning a selected sample of individuals to obtain data about a process, product, or service (ASQ QP Glossary).

synchronous and asynchronous—In a synchronous system, operations are coordinated under the centralized control of a fixed rate. An asynchronous system, in contrast, has no fixed rate but instead operates under distributed control. *Active* (live, real-time) communications for the exchange or dissemination of information would be *synchronous*; communication is time and/ or place-dependent. In most cases, you must be at a specific place at a specific time. On the other hand, *passive* (reading, viewing media, listening to a recording) communications for the exchange or dissemination of information would be *asynchronous*; Communication is access and availability (distribution) dependent. You don't need to be at a specific place at a specific time.

telecommuter—A person who works at home using an electronic linkup with a central office.

telecommuter workforce—Full- or part-time employees who work from the road or a home office. Many companies are replacing permanent offices with temporary workstations for day use for telecommuters visiting the business location.

video call—Use of cameras and audio so parties can see and talk to one another.

virtual—Being on or simulated on a computer or computer network <print or *virtual* books> <a *virtual* keyboard>: as a: occurring or existing primarily online <a *virtual* library> <*virtual* shopping> b: of, relating to, or existing within a virtual reality <a *virtual* world> <a *virtual* tour>. Accessed August 5, 2024, http://unabridged. Merriam-webster.com/.

virtual meeting—A meeting using electronic means to exchange information or respond to questions as if you were in a meeting room or with the interviewee at his or her workstation.

virtual organization or company—One with members who are geographically apart, usually working by email and groupware, while appearing to outsiders to be a single, unified organization with a real physical location. A *virtual corporation* (VC) is a group of independent companies that functions as one entity through telecommunications and computer technology. Virtual corporations' strengths include agility and resourcefulness in obtaining subject matter experts. VCs can be short-term, such as for a summer concert tour, or a long-term business strategy that sets up OEM partnerships. An example is Boeing's partnership with General Electric, which engineers and manufactures Boeing 787 engines.

webinars—A lecture using electronic means to disseminate information. It may be live or pre-recorded, and there may be questions at the end.

References

1. ASQ Quality Glossary. https://asq.org/quality-resources/quality-glossary.

2. Hammond, Sue Annis. *The Thin Book of Appreciative Inquiry.* Plano, TX: Thin Book Publishing Co., 1996.

3. Lipnack, Jessica, and Jeffrey Stamps. *Virtual Teams: Reaching across Space, Time, and Organizations with Technology.* New York: John Wiley & Sons, 1997.

4. Littlejohn, Stephen W. *Theories of Human Communication,* Seventh Edition. Belmont, CA: Wadsworth/Thomas Learning, 2002.

5. Russell, J.P. "Remote Control: Is e-Auditing the Inevitable Next Step?" *Quality Progress,* January 2011.

6. Wilson, Shauna. "Forming Virtual Teams." *Quality Progress,* June 2003.

7. Wilson, Shauna. *InterneTeaming.com: Tools to Create High-Performance Remote Teams.* Portland, OR: Inkwater Publishing, 2005.

8. Wilson, Shauna. "Auditing in Virtual Environments." *The Auditor,* March–April 2008.

9. Wilson, Shauna. "Successful Online Meeting InfoLine." *ASTD Publishing,* Issue 0902, February 2009.

10. Wilson, Shauna. 2013. "Auditing Communication in a Global Supply Chain." *The Auditor,* Jan-Feb 2013.

11. Wilson, Shauna. "Setting the Foundation for Conducting eAudits." *The Auditor,* July 2016.

12. Wilson, Shauna. "eAuditing: A matter of context." *The Auditor,* March 2017.

About the Authors

Shauna Wilson

Having worked in quality for over forty years, Shauna Wilson's experience in virtual team communication and development of virtual auditing methods has made her a leading expert in remote auditing. She earned an MS in Engineering focused on Organizational Performance Technologies and Instructional Design. Wilson wrote *InterneTeaming. com: Tools to Create High Performance Remote Teams* and co-authored *eAuditing Fundamentals: Virtual Communication and Remote Auditing.* She has been featured in *Quality Progress, The Auditor,* and ASTD's *InfoLine,* and she has served as the US Expert for PC/TAG302 ISO19011:2018 Auditing Management Systems Guideline. Wilson is the 2022 recipient of the ASQ Audit Division Paul Gauthier Award and holds the Certus Master Auditor certification.

Paul Russell

Paul Russell has performed remote audits with two US Defense Agency contracts. He is Managing Director for QualityWBT Center for Education and the JP Russell Learning Center. For 21 years, Paul Russell has worked with Web-based training and blended learning classes approved by the American Society for Quality. He is a voting member of the United States Technical Advisory Group (US TAG) ISO/TC 176 for Quality Management Systems standards. Previously, Paul was in the health industry maintaining Food and Drug Administration (FDA) and European Union (EU) compliance for over 45 blood bank locations, five blood manufacturing facilities, and over 31 blood mobiles. While performing pharmacy audits, he was assigned the Midwest/Northeast/Mid-Atlantic/South regions of the US to maintain Drug Enforcement Agency (DEA) and FDA compliance as well as pharmacy specific adherence to the drug laws of 21 states and one District.